Encyclopaedia of
World Geography

Encyclopaedia of World Geography

Vol. - 5

Chandra Vijay Purty

ABD PUBLISHERS
Jaipur ■ New Delhi

ABD PUBLISHERS

Regd. Off. "Bony Residency", Gate No. 2
Opp. Tilak Public School, Vishweswariya Nagar,
Gopalpura Road, Jaipur - 302018, India
Ph: 0141 - 2761280, Telefax: 2761381
Email: abdpublishersjpr@gmail.com

Delhi Office: 102, 1st Floor, Satyam House,
4327/3, Ansari Road, Darya Ganj,
New Delhi - 110002, India
Ph: 011-45652440

visit us: www.oxfordbookcompany.com

Published in India, 2024

ISBN : 9788183763127 (7 Vols. Set)

Typeset by Shivangi Computer, Jaipur

Printed and bound in India by Thomson Press Ltd.,
New Delhi

Preface

Geography is the study of the Earth and its lands, features, inhabitants, and phenomena. A literal translation would be "to describe or write about the Earth". The first person to use the word "geography" was Eratosthenes (276-194 B.C.). Four historical traditions in geographical research are the spatial analysis of natural and human phenomena (geography as a study of distribution), area studies (places and regions), study of man-land relationship, and research in earth sciences. Nonetheless, modern geography is an all-encompassing discipline that foremost seeks to understand the Earth and all of its human and natural complexities—not merely where objects are, but how they have changed and come to be. As "the bridge between the human and physical sciences," geography is divided into two main branches—human geography and physical geography.

The oldest known world maps date back to ancient Babylon from the 9th century BC. The best known Babylonian world map, however, is the *Imago Mundi* of 600 BC. The map as reconstructed by Eckhard Unger shows Babylon on the Euphrates, surrounded by a circular landmass showing Assyria, Urartu and several cities, in turn surrounded by a "bitter river" (Oceanus), with seven islands arranged around it so as to form a seven-pointed star. The accompanying text mentions seven outer regions beyond the encircling ocean. The descriptions of five of them have survived. In contrast to the *Imago Mundi*, an earlier Babylonian world map dating back to the 9th century BC depicted Babylon as being further north from the centre of the world, though it is not certain what that centre was supposed to represent.

The ideas of Anaximander (c. 610 B.C.-c. 545 B.C.), considered by later Greek writers to be the true founder of geography, come to us through fragments quoted by his successors. Anaximander is credited with the invention of the gnomon, the simple yet efficient Greek instrument that allowed the early measurement of latitude. Thales, Anaximander is also credited with the prediction of eclipses. The foundations of geography can be traced to the ancient cultures, such as the ancient, medieval, and early modern Chinese. The Greeks, who were the first to explore geography as both art and science, achieved this through Cartography, Philosophy, and Literature, or through Mathematics. There is some debate about who was the first person to assert that the Earth is spherical in shape, with the credit going either to Parmenides or Pythagoras. Anaxagoras was able to demonstrate that the profile of the Earth was circular by explaining eclipses. However, he still believed that the Earth was a flat disk, as did many of his contemporaries. One of the first estimates of the radius of the Earth was made by Eratosthenes.

The first rigorous system of latitude and longitude lines is credited to Hipparchus. He employed a sexagesimal system that was derived from Babylonian mathematics. The parallels and meridians were sub-divided into 360°, with each degree further subdivided 602 (minutes). To measure the longitude at different location on Earth, he suggested using eclipses to determine the relative difference in time. The extensive mapping by the Romans as they explored new lands would later provide a high level of information for Ptolemy to construct detailed atlases. He extended the work of Hipparchus, using a grid system on his maps and adopting a length of 56.5 miles for a degree.

From the 3rd century onwards, Chinese methods of geographical study and writing of geographical literature became much more complex than what was found in Europe at the time (until the 13th century). Chinese geographers such as Liu An, Pei Xiu, Jia Dan, Shen Kuo, Fan Chengda, Zhou Daguan, and Xu Xiake wrote important treatises, yet by the 17th century, advanced ideas and methods of Western-style geography were adopted in China.

This book contains the fundamental and basic information of the subject.

Chandra Vijay Purty

Contents

1

Suriname

Suriname, officially the Republic of Suriname, is a country in northern South America.

Suriname is situated between French Guiana to the east and Guyana to the west. The southern border is shared with Brazil and the northern border is the Atlantic coast. The southernmost borders with French Guiana and Guyana are disputed along the Marowijne and Corantijn rivers, respectively; while a part of the disputed maritime boundary with Guyana was arbitrated by the United Nations Convention on Law of the Sea on September 20, 2007.

Suriname, at 163,000 sq. km is the smallest sovereign state in terms of area in South America. The country is the only Dutch-speaking region in the world that is not a part of the Kingdom of the Netherlands or Belgium, and is the only state outside Europe with Dutch as an official language. The combined legacy of years of colonial occupation, immigration, and slavery has made Suriname one of the most multicultural societies in the world, with extreme ethnic, linguistic, and religious diversity. Suriname's geographical size is just under 165,000 km^2 (64,000 sq mi), and it has an estimated population of about 470,000 people, most of whom live on the country's north coast.

Etymology

The name, Suriname, may derive from a Taino (Arawak-speaking) group called "Surinen" who first inhabited the region prior to European arrival. Originally, the country was spelled *Surinam* by English settlers who founded the first colony at

Marshall's Creek, along the Suriname River, and was formerly known as *Nederlands Guyana*, *Netherlands Guiana* or *Dutch Guiana*. "Surinam" can still be found in English. A notable example of this is Suriname's own national airline, Surinam Airways.

Geography

Suriname is the smallest independent country in South America. Situated on the Guiana Shield, the country can be divided into two main geographic regions. The northern, lowland coastal area (roughly above the line Albina-Paranam-Wageningen) has been cultivated, and most of the population lives here.

The southern part consists of tropical rainforest and sparsely inhabited savanna along the border with Brazil, covering about 80% of Suriname's land surface.

There are two main mountain ranges in the Bakhuys Mountains and the Van Asch Van Wijck Mountains. Julianatop is the highest mountain in the country at 1,286 metres (4,219 ft) above sea level.

Other mountains include Tafelberg at 1,026 metres (3,366 ft), Mount Kasikasima at 718 metres (2,356 ft), Goliathberg at 358 metres (1,175 ft) and Voltzberg at 240 metres (790 ft).

Suriname is located in northern South America and is part of Caribbean South America, bordering the North Atlantic Ocean, between French Guiana and Guyana. It is mostly covered by tropical rain forest, containing a great diversity of flora and fauna that, for the most part, are increasingly threatened by new development. There is a relatively small population, most of which live along the coast.

Location

Area

Total: 163,270 square kilometers (63,040 sq mi)

Land: 161,470 square kilometers (62,340 sq mi)

Water: 1,800 square kilometers (690 sq mi).

Area - comparative: See order of magnitude 1 E+11 m^2.

Land Boundaries

Total: 1,707 kilometers (1,061 mi)

Border countries:

- Brazil - 597 kilometers (371 mi)
- French Guiana - 510 kilometers (317 mi)
- Guyana - 600 kilometers (373 mi).

Coastline: 386 kilometers (240 mi).

Maritime Claims

Exclusive economic zone: 200 nmi (370.4 km; 230.2 mi).

Territorial sea: 12 nmi (22.2 km; 13.8 mi).

Climate and Terrain

Climate

Suriname has a tropical environment, and is moderated by strong winds.

Terrain

Most of the country is made up of rolling hills, but there is a narrow coastal plain that has swampy terrain.

Elevation Extremes

Lowest point: Unnamed location in the coastal plain - 2 meters (7 ft) below Sea Level.

Highest point: Juliana Top - 1,280 meters (4,200 ft)

Natural Resources

Timber, hydropower, fish, kaolin, shrimp, bauxite and gold. Small amounts of nickel, copper, platinum and iron ore.

Water

The country has one large reservoir, the Brokopondo Reservoir. Several rivers run through it, including the Suriname River, Nickerie River and Maroni or Marowijne River.

Land Use

(Using 1993 Estimates)

Arable land: .36%

Permanent crops: >0.6%

Permanent pastures: 0%

Other: 99.58%

Irrigated Land

(Using 1993 Estimates)

600 square kilometers (232 sq mi).

Natural hazards: N/A.

Environment

Current Issues

Deforestation is a real problem as timber is cut for export. There is also a lot of pollution of inland waterways by small-scale mining activities.

Districts and Resorts

Suriname is divided into ten districts:

- Brokopondo
- Commewijne
- Coronie
- Marowijne
- Nickerie
- Para
- Paramaribo
- Saramacca
- Sipaliwini
- Wanica.

Suriname is further subdivided into 62 resorts (*ressorten*).

Climate

Lying 2 to 5 degrees north of the equator, Suriname has a very hot tropical climate, and temperatures do not vary much throughout the year. The year has two wet seasons, from April to August and from November to February. It also has two dry seasons, from August to November and February to April.

Nature Reserves

In the upper Coppename River watershed, the Central Suriname Nature Reserve is a UNESCO World Heritage Site cited for its unspoiled rainforest biodiversity. There are many national parks in the country: Galibi National Reserve, Coppename Manding National Park and Wia Wia NR along the coast, Brownsberg NR, Raleighvallen/Voltzeberg NR, Tafelberg NR and Eilerts de Haan NP in the centre and the Sipaliwani NR on the Brazilian border. In all, 12.6% of the country's land area are national parks and lakes, according to the UNEP World Conservation Monitoring Centre.

History

Beginning in the 16th century, the area was discovered by French, Spanish, and English explorers. A century later, plantation colonies were established by the Dutch and English along the many rivers in the fertile Guyana plains. The earliest documented colony in Guiana was along the Suriname River and called Marshall's Creek. The area was named after an Englishman. Disputes arose between the Dutch and the English. In 1667, the Dutch decided to keep the nascent plantation colony of Suriname conquered from the English, resulting from the Treaty of Breda. The English were left with New Amsterdam, a small trading post in North America, which later became New York City.

The planters relied heavily on African slaves to cultivate the coffee, cocoa, sugar cane and cotton plantations along the rivers. Treatment of the slaves by their owners was notoriously bad, and many slaves escaped the plantations. With the help of the native South Americans living in the adjoining rain forests, these runaway slaves established a new and unique culture that was highly successful in its own right. Known collectively in English as the Maroons, in French as the Neg Marrons and in Dutch as "Bosnegers" (literally meaning "bush negroes"), they actually established several independent tribes, among them the Saramaka, the Paramaka, the Ndyuka or Aukan, the Kwinti, the Aluku or Boni, and the Matawai.

The Maroons would often raid the plantations to recruit new members, acquire women, weapons, food and supplies. These attacks were often deadly for the planters and their families, and after several unsuccessful campaigns against the Maroons, the European authorities signed several peace treaties with them in the 19th century, granting the Maroons sovereign status and trade rights. Slavery was abolished by the Netherlands in Suriname in 1863, but the slaves in Suriname were not fully released until 1873, after a mandatory 10 year transition period during which time they were required to work on the plantations for minimal pay and without state sanctioned torture. As soon as they became truly free, the slaves largely abandoned the plantations where they had suffered for several generations, in favour of the city, Paramaribo. As a plantation colony, Suriname was still heavily dependent on manual labour, and to make up for the shortfall, the Dutch brought in contract labourers from the Dutch East Indies (modern Indonesia) and India (through an arrangement with the British). In addition, during the late 19th and early 20th centuries, small numbers of mostly men were brought in from China and the Middle East. Although Suriname's population remains relatively small, because of this history it is one of the most ethnically and culturally diverse in the world.

In 1954, the Dutch placed Suriname under a system of limited self-government, with the Netherlands retaining control of defense and foreign affairs. In 1973, the local government, led by the NPK (a largely Creole, meaning ethnically African or mixed African-European, party) started negotiations with the Dutch government leading towards full independence, which was granted on 25 November 1975. The severance package was very substantial, and a large part of Suriname's economy for the first decade following independence was fueled by foreign aid provided by the Dutch government.

The first President of the country was Johan Ferrier, the former governor, with Henck Arron (the then leader of the Nationale Partij Suriname (Suriname's National Party)) as Prime Minister. Nearly one third of the population of Suriname at that time emigrated to the Netherlands in the years leading up to independence, as many people feared that the new country would fare worse under independence than it did as an overseas colony of the Netherlands. Suriname's diaspora therefore includes more than a quarter of one million people of Suriname origin living in the Netherlands today, including several recent members of the Dutch national football (soccer) team.

On February 25, 1980, a military coup sidelined the democratic government and declared a Socialist Republic, and with it began a period of economic and social hardship for the country. On 8 December 1982, the military, then under the leadership of Desi Bouterse, rounded up several prominent citizens who were accused of plotting against the government. They were allegedly tortured and certainly killed during the night, and the Netherlands quickly suspended all foreign aid to Suriname after this event. (As of August 2008, Desi Bouterse is currently standing trial in Suriname for his role in these killings.)

Elections were held in 1987 and a new constitution was adopted, which among other things allowed the dictator to remain in charge of the army. Dissatisfied with the government, Bouterse summarily dismissed them in 1990, by telephone. This event became popularly known as "the telephone coup". Bouterse's power began to wane after the 1991 elections however, and a brutal civil war between the Suriname army and the Maroons, loyal to the rebel leader Ronnie Brunswijk, further weakened his position during the 1990s.

Suriname's democracy gained some strength after the turbulent 1990s, and its economy became more diversified and less dependent on Dutch financial assistance. Bauxite (Aluminum ore) mining continues to be a strong revenue source, but the discovery and exploitation of oil and gold has added substantially to Suriname's economic independence. Agriculture, especially of rice and bananas, remains a strong component of the economy, and ecotourism is providing new economic opportunities. More than 80% of Suriname's land-mass consists of unspoiled rain forest, and with the establishment of the Central Suriname Nature Reserve in 1998, Suriname signaled its commitment to conservation of this precious resource. The Central Suriname Nature Reserve became a World Heritage Site in 2000.

Economy

The economy of Suriname is dominated by the bauxite industry, which accounts for more than 15% of GDP and 70% of export earnings. Other main export products include rice, bananas and shrimp. Suriname has recently started exploiting some of its sizeable oil and gold reserves. About a quarter of the people work in the agricultural sector. The Surinamese economy is very dependent on commerce, its main trade partners being the Netherlands, the United States, Canada and Caribbean countries. After assuming power in the fall of 1996, the Wijdenbosch government ended the structural adjustment program of the previous government, claiming it was unfair to the poorer elements of society. Tax revenues fell as old taxes lapsed and the government failed to implement new tax alternatives. By the end of 1997, the allocation of new Dutch development funds was frozen as Surinamese Government relations with the Netherlands deteriorated. Economic growth slowed in 1998, with decline in the mining, construction, and utility sectors. Rampant government expenditures, poor tax collection, a bloated civil service, and reduced foreign aid in 1999 contributed to the fiscal deficit, estimated at 11% of GDP. The government sought to cover this deficit through monetary expansion, which led to a dramatic increase in inflation.

GDP (2006 est.): U.S. $2.11 billion.

Annual growth rate real GDP (2006 est.): 5.8%.

Per capita GDP (2006 est.): U.S. $4,000.

Inflation (2006): 5.6%.

Natural resources: Bauxite, gold, oil, iron ore, other minerals; forests; hydroelectric potential; fish and shrimp.

Agriculture: Products—rice, bananas, timber, and citrus fruits.

Industry: Types—alumina, oil, gold, fish, shrimp, lumber.

Trade (2005):

Exports—U.S. $929.1 million: alumina, gold, crude oil, wood and wood products, rice, bananas, fish, and shrimp. Major markets—Norway (23.9%), U.S. (16.8%), Canada (16.4%), France (8.1%), Iceland (2.9%).

Imports—$1.1 billion: capital equipment, petroleum, iron and steel products, agricultural products, and consumer goods. Major suppliers—U.S. (24.4%), Netherlands (14.5%), Trinidad and Tobago (10.5%), China (5.4%), Japan (4.3%), Brazil (3.6%).

Demographics

In November 2007, Suriname's population was estimated to be 494,347. It is made up of several distinct ethnic groups.

- Hindoestanen form the largest major group at 27% of the population. They are descendants of nineteenth-century contract workers from India. They are from the Indian states of Bihar and Eastern Uttar Pradesh, in Northern India, along the Nepali border.
- The Surinamese Creoles form the middle group 18% of the population. They are the mixed descendants of West African slaves and Europeans (mostly Dutch).
- The Javanese (descendants of contract workers from the former Dutch East Indies on the island of Java, Indonesia) make up 15% (close to 90,000) of the population.
- Surinamese Maroons (descendants of escaped West African slaves) make up 15% and are divided into five main groups: Ndyuka (Aucans), Kwinti, Matawai, Saramaccans and Paramaccans.
- Mixed 12.5%
- Amerindians form 3.7% of the population, the main groups being the Akuriyo, Arawak, Carib/Kalina, Trio (Tiriyo), and Wayana.
- Chinese, mainly descendants of the earliest nineteenth-century contract workers, 1.8% and number about 14,000.
- Boeroes (derived from *boer*, the Dutch word for *farmer*) are descendants of Dutch nineteenth-century immigrant farmers. Most Boeroes left after independence in 1975.
- Brazilians, many of them gold miners. Most of the nearly 40,000 Brazilians living in Suriname arrived during the past several years.

There is no predominant religion in the country. Christianity, both in the form of Roman Catholicism and various denominations of Protestantism, is dominant among Creoles and Maroons. The Creoles and to a lesser degree the Maroons, both descendants of enslaved Africans, were forced to convert to Christianity, but a lot of them still retain their Afro-American religion called Winti. Most of the Hindustani are Hindu, but some practice Islam or Christianity. The Javanese practice either Islam or Christianity. Suriname's population is 20% Muslim, which is the highest minority-percentage of Muslims of any country in the New World. The makeup of Suriname's population is similar to that of neighboring Guyana, with the exception of the large Indonesian population (which is not present in Guyana). French Guiana, which is a part of France, does not collect ethnic statistics, but is believed to contain much smaller Hindustani and Indonesian populations.

The vast majority of people (about 90%) live in Paramaribo or on the coast. There is also a significant Surinamese population in the Netherlands. In 2005 there were 328,300 Surinamese people living in the Netherlands, which is about 2% of the total population of the Netherlands, compared to 438,000 Surinamese in Suriname itself. These groups are all represented in the old flag of Suriname.

Languages

Surinamese society is one of the most multilingual in the world. Dutch is the sole official language, and is the language of education, government, business and the media. Over 60 percent of the population speak it as a mother tongue, and most of the rest speak it as a second or third language. In 2004, Suriname became an associate member of the Dutch Language Union. It is the only Dutch-speaking country in South America.

In Paramaribo, Dutch is the main home language in two-thirds of households. The recognition of *"Surinaams Nederlands"* (*"Surinam Dutch"*) as a natiolect equal to *"Nederlands Nederlands"* (*"Dutch Dutch"*) and *"Vlaams Nederlands"* (*"Flemish Dutch"*) was expressed in 2009 by the publication of the *Woordenboek Surinaams Nederlands* (*Surinamese Dutch Dictionary*). Only in the interior of Suriname is Dutch seldom used.

Sranan Tongo, a local creole language originally spoken by the Creole population group, is the most widely used language in the streets and often interchangeably with Dutch depending on the formality of the setting.

Surinamese Hindi or Sarnami, a dialect of Bhojpuri, is the third-most used language, spoken by the descendants of British Asian contract workers. Javanese is used by the descendants of Javanese contract workers. The Maroon languages, somewhat intelligible with Sranan Tongo, include Saramaka, Paramakan, Ndyuka, *Aukan*, Kwinti and Matawai. Amerindian languages, spoken by Amerindians, include Carib and Arawak. Hakka and Cantonese are spoken by the descendants of the Chinese contract (*koelie*, coolie) workers. Mandarin is spoken by some few recent Chinese immigrants. English, Spanish and Portuguese are also used. Spanish and Portuguese are spoken by Latin American residents and their descendants and sometimes also taught in schools.

The public discourse about Suriname's languages is a part of an ongoing debate about the country's national identity. While Dutch is perceived as a remnant of colonialism by some, the use of the popular Sranan became associated with nationalist politics after its public use by former dictator Dési Bouterse in the 1980s, and groups descended from escaped slaves might resent it. Some propose to change the national language to English, so as to improve links to the Caribbean and North America, or to Spanish, as a nod to Suriname's location in South America, although it has no Spanish-speaking neighbours.

Health

Fertility rate was at 2.6 births per woman. Public expenditure was at 3.6 % of the GDP in 2004, whereas private expenditure was at 4.2 %. There were 45 physicians per 100,000 in the early 2000s. Infant mortality was at 30 per 1,000 live births. Male life expectancy at birth was at 66.4 years, whereas female life expectancy at birth was at 73 years.

Transport

Suriname and neighboring Guyana are the only two countries on the (in-land) American continent who still drive on the left. In Guyana this practice is inherited from United Kingdom colonial authorities. The reason for the left hand drive in Suriname is explained by several sources. It is thought that this is because the first cars imported were from England, but this is yet undocumented. In addition, this view does not make statements on traffic before the automobile era. Another explanation is that the Netherlands, at the time of colonization of Suriname, used the left-hand side of the road for traffic, or that Suriname was first colonized by the English. Although the Netherlands converted to driving to the right at the end of the 18th century (Peter Kincaid, and), Suriname did not. An interesting viewpoint on this is forwarded by Peter Kincaid, and further explored by Ian Watson, in that territories such as Suriname, with no neighbours or no connecting roads to neighbour countries, had no external pressure to either change or to maintain the sta us quo on driving sides.

Politics

The Republic of Suriname is a constitutional, democratically representational republic based on the 1987 constitution. The legislative branch of government consists of a 51-member unicameral National Assembly, simultaneously and popularly elected for a five-year term. During the recently held elections on Tuesday 25th of May the "Megacombinatie" won 23 of the National Assembly seats followed by "Nationale Front" with 20 seats. A much smaller but important for the collaboration went to the 'A-combinatie" and to the "Volksalliantie". For more details the reader can visit the website of the "25 mei verkiezingen.com" or from the "ministerie van binnenlandse zaken" which is the ministry of internal affairs. Currently negiotations are going on in and between parties regarding the formation of the coalition for the coming five years.

The president, who is elected for a five-year term by a two-thirds majority of the National Assembly or, failing that, by a majority of the People's Assembly, heads the executive branch. If at least two-thirds of the National Assembly cannot agree to vote for one presidential candidate, a People's Assembly is formed from all National Assembly delegates and regional and municipal representatives who were elected by popular vote in the most recent national election. As head of government, the president appoints a 16-minister cabinet. A vice president, normally elected at the same time as the president, needs a simple majority in the National Assembly or People's Assembly to be elected for a 5-year term. As head of government, the president appoints a cabinet of ministers. There is no constitutional provision for removal or replacement of the president unless he resigns.

The judiciary is headed by the Court of Justice (Supreme Court). This court supervises the magistrate courts. Members are appointed for life by the president in consultation with the National Assembly, the State Advisory Council and the National Order of Private Attorneys. In April 2005, the regional Caribbean

Court of Justice, based in Trinidad, was inaugurated. As the final court of appeal, it was intended to replace the London-based Privy Council.

The country is divided into 10 administrative districts, each headed by a district commissioner appointed by the president. The commissioner is similar to the governor of a United States-type state, but is appointed and removed by the president.

Culture

Owing to the country's multicultural heritage, Suriname celebrates a variety of distinct ethnic and religious festivals.

National Celebrations

January 1 - New Year's Day

March 1 (varies) - Holi/ Phagwa

May 1 - Labour Day

June 5 - Immigration of the Indians

July 1 - Keti Koti, Emancipation Day (end of slavery)

August 8 - Day of the indigenous people

August 9 - Immigration of the Javanese

November 25 - Independence Day

December 5 - Children's day (Sinterklaas)

December 25 - Christmas Day

December 26 - Second Christmas Day.

There are several Hindu and Islamic national holidays like Divali (deepavali), Phagwa and Eid ul-Fitr and Eiduladha. These holidays do not have specific dates on the Gregorian calendar as they are based on the Hindu and Islamic calendars, respectively.

There are several holidays which are unique to Suriname. These include the Indian, Javanese and Chinese arrival days. They celebrate the arrival of the first ships with their respective immigrants.

New Year's Eve

New Year's Eve in Suriname is called *Oud jaar*, or "old year". It is during this period that the Surinamese population goes to the city's commercial district to watch demonstrational fireworks. The bigger stores invest in these firecrackers and display them out in the streets. Every year the length of them is compared, and high praises are held for the company that has managed to import the largest ribbon.

These celebrations start at 10 in the morning and finish the next day. The day is usually filled with laughter, dance, music, and drinking. When the night starts, the big street parties are already at full capacity. The most popular fiesta is the one that is held at cafe 't Vat in the main tourist district. The parties there stop between 10 and 11 at night. After which the people go home to light their pagaras (red-firecracker-ribbons) at midnight. After 12, the parties continue and the streets fill again until daybreak.

Sports

Some of the greatest football players to represent the Netherlands, such as Pierre van Hooijdonk, Frank Rijkaard, Ruud Gullit, Patrick Kluivert, Edgar Davids, Clarence Seedorf, Aron Winter, Jimmy Floyd Hasselbaink, Stanley Menzo, Ryan Babel, Ken Monkou, Edson Braafheid, Boy Waterman, Regi Blinker and Fabian Wilnis are of Surinamese descent. Davids in particular has written of his passionate pride in his Surinamese heritage and his love of attending football matches there. There are a number of local heroes in other sports as well, like Primraj Binda, best known as the athlete who dominated the local 10 km for nearly a decade, Steven Vismale and Letitia Vriesde. Another notable track athlete from Suriname was Tommy Asinga.

Anthony Nesty is the only person to win a medal (for swimming) for Suriname at the Olympics. Originally from Trinidad and Tobago, not Suriname, he now lives in Gainesville, Florida, USA, and is a coach of the University of Florida. He is mainly a distance coach.

Multiple K-1 champion and legend, Ernesto Hoost, was born in Suriname. Remy Bonjasky also a multiple K-1 champion is also from Surinamese descent. MMA and Kickboxing champions Melvin Manhoef, Gilbert Yvel and Alistair Overeem were born in Suriname or from Surinamese descent. Retired female kickboxer Ilonka Elmont was also born in Suriname.

Another notable up and comer kickboxer and K-1 fighter, Tyrone Spong, was born in Suriname. Ginty Vrede, a former Muay Thai Heavy Weight Champion who passed away in 2008 aged 22, was born in Suriname.

Education

The net primary enrollment rate was 94 % in 2004. Education is compulsory until the age of 12. Literacy is very common, particularly among males. The university of the country is the Anton de Kom University of Suriname.

Media

A popular newspaper is *De Ware Tijd*. Suriname has 24 radio stations from which a couple broadcast through the Internet (Apintie and Radio10). There are also a dozen television networks including STVS, RBN, ABC, ATV, Mustika, and Garuda). Also listened to is mArt, a broadcaster from Amsterdam founded by people from Suriname. Kondreman is one of the popular cartoons in Suriname.

Tourism

Royal Torarica, was opened in the night district of Paramaribo on the Suriname River. The hotel industry is important to Suriname's economy. The rental of apartments, or the rent-a-house phenomenon, is also popular in Suriname.

Most tourists visit Suriname for the outstanding biodiversity of the pristine Amazonian rain forests in the south of the country, which are noted for their flora and fauna. The Central Suriname Nature Reserve is the biggest and one of the most popular reserves, along with the Brownsberg Nature Park which overlooks the Brokopondo Reservoir, the latter being one of the largest man-made lakes in the world. Tonka Island in the reservoir is home to a rustic eco-tourism project run by the Saramaccaner Maroons. There are also many waterfalls throughout the country: Raleighvallen, or Raleigh Falls, is a 56,000 hectare nature reserve on the Coppename River, rich in bird life. Also are the Blanche Marie Falls on the Nickerie River and the Wonotobo Falls. Tafelberg Mountain in the centre of the country is surrounded by its own reserve- the Tafelberg Nature Reserve- around the source of the Saramacca River, as is the Voltzberg Nature Reserve further north on the Coppename River at Raleighvallen. In the interior are many Maroon and Amerindian villages which often have their own reserves and are open to visitors.

Suriname is one of the few countries in the world where at least one of each biome that the state possesses has been declared a wildlife reserve. Around 30% of the total land area of Suriname is protected by law as reserves.

Other attractions include plantations such as Laarwijk, which is situated along the Suriname River. This plantation can only be reached by boat via Domburg, in the north central Wanica District of Suriname.

Landmarks

The Jules Wijdenbosch Bridge is a bridge over the river Suriname between Paramaribo and Meerzorg in the Commewijne district. The bridge was built during the tenure of President Jules Albert Wijdenbosch (1996–2000) and was completed in 2000. The bridge is 52 metres (171 ft) high, and 1,504 metres (4,934 ft) long. It connects Paramaribo with Commewijne, a connection which previously could only be made by ferry. The purpose of the bridge was to facilitate and promote the development of the eastern part of Suriname. The bridge consists of two lanes and is not accessible to pedestrians.

The Cathedral of St. Peter and Paul is 114 years old. Before it became a cathedral it was a theatre and was owned by La Parra. The theatre was built in 1809 and burned down in 1820. The construction of the Sts. Peter and Paul Cathedral started on January, 13, 1883.

Suriname is one of the few countries in the world where a synagogue is located next to a mosque (another place is Sofia, Bulgaria). The two buildings are located next to each other in the centre of Paramaribo and have been known to share a parking facility during their respective religious rites, should they happen to coincide with one another.

2

French Guiana

French Guiana is an overseas region of France, consisting of a single overseas department (French: *departement d'outre-mer, or DOM*) located on the northern coast of South America. It has borders with two nations, Brazil and Suriname. Its currency is the euro. Almost half the population lives in the suburbs surrounding the region's main city, Cayenne.

The addition of the adjective "French" comes from colonial times when three such colonies existed: British Guiana (now Guyana), Dutch Guiana (now Suriname) and French Guiana. The three are still often collectively referred to as the Guianas.

History

French Guiana was originally inhabited by a number of indigenous American people. It was settled by the French during the 17th century. After the Treaty of Paris in 1763, Louis XV sent 12,000 settlers to French Guiana to colonise the region. One and a half years later only a few hundred survived. Its infamous *Île du Diable* (Devil's Island) was the site of penal settlements from 1852 until 1951. More than 70,000 French convicts were deported to French Guiana between 1852 and 1939.

In 1809 a Portuguese-British naval squadron took French Guiana for the Portuguese Empire. With the signing of the Treaty of Paris in 1814 the region was handed back to the French, though a Portuguese presence remained until 1817.

A border dispute with Brazil arose in the late 19th century over a vast area of jungle, leading to the short-lived pro-French independent state of Counani in the disputed territory and some fighting between settlers, before the dispute was resolved largely in favour of Brazil by the arbitration of the Swiss government.

In 1946, French Guiana became an overseas department of France. The 1970s saw the settlement of Hmong refugees from Laos. A movement for increased autonomy from France gained some momentum in the 1970s and 1980s, but has since abated.

Geography

Though sharing cultural affinities with the French-speaking territories of the Caribbean, French Guiana is not considered to be part of that geographic region, because the Caribbean Sea is actually located several hundred kilometres to the west, beyond the arc of the Lesser Antilles.

French Guiana consists of two main geographical regions: a coastal strip where the majority of the people live, and dense, near-inaccessible rainforest which gradually rises to the modest peaks of the Tumac-Humac mountains along the Brazilian frontier. French Guiana's highest peak is Bellevue de l'Inini (851 m). Other mountains include Mont Machalou (782 m), Pic Coudreau (711 m) and Mont St Marcel (635 m), Mont Favard (200 m) and Montagne du Mahury (156 m). Several small islands are found off the coast, the three Iles du Salut Salvation Islands which includes Devil's Island and the isolated Iles du Connétable bird sanctuary further along the coast towards Brazil. The Barrage de Petit-Saut hydroelectric dam in the north of French Guiana forms an artificial lake and provides hydroelectricity. There are many rivers in French Guiana.

As of 2007, the Amazonian forest located in the most remote part of the department is now protected through one of the nine national parks of France, the Guiana Amazonian Park. The territory of the park covers some 33,900 square kilometres (13,090 sq mi) upon the communes of Camopi, Maripasoula, Papaïchton, Saint-Élie and Saül.

Economy

French Guiana is heavily dependent on mainland France for subsidies, trade, and goods. The main industries are fishing (accounting for three-quarters of foreign exports), gold mining and timber. In addition, the Guiana Space Centre at Kourou accounts for 25% of the GDP and employs about 1,700 people.

There is very little manufacturing. Agriculture is largely undeveloped and is mainly confined to the area near the coast — sugar and bananas are two of the main cash crops grown. Tourism, especially eco-tourism, is growing. Unemployment is a major problem, running at about 20% to 30%.

In 2006 the GDP per capita of French Guiana at market exchange rates, not at PPP, was €13,800 (US$17,380), which was 48% of Metropolitan France's average GDP per capita that year.

Demographics

French Guiana's population of 221,500 (January 2008 est.), most of whom live along the coast, is very ethnically diverse. At the 1999 census, 54.4% of the inhabitants of French Guiana were born in French Guiana, 11.8% were born in Metropolitan France, 5.2% were born in the French Caribbean *departements* (Guadeloupe and Martinique), and 28.6% were born in foreign countries (primarily Brazil, Suriname, and Haiti).

Estimates of the percentages of French Guiana ethnic composition vary, a situation compounded by the large proportion of immigrants (about 20,000, nearly 10%).

Creoles (people of mixed African and French ancestry) are the largest ethnic group, though estimates vary as to the exact percentage, depending upon whether the large Haitian community is included as well. Generally the Creole population is judged to be about 60% to 70% of the total population if Haitians (comprising roughly one-third of Creoles) are included, and 30% to 50% without.

Roughly 14% of the population is of European ancestry. The vast majority of these are of French heritage, though there are also people of Dutch, British, Spanish and Portuguese ancestry .

The main Asian communities are the Chinese (3.2%, primarily from Hong Kong and Zhejiang province) and Hmong from Laos (1.5%). There are also smaller groups from various Caribbean islands, mainly Saint Lucia as well as Dominica.

Other Asian groups include East Indians, Lebanese and Vietnamese.

The main groups living in the interior are the Maroons (formerly called "Bush Negroes") who are racially black African, and Amerindians. The Maroons, descendants of escaped African slaves, live primarily along the Maroni River. The main Maroon groups are the Saramaca, Aucan (both of whom also live in Suriname), and Boni (Aluku).

The main Amerindian groups (forming about 3%-4% of the population) are the Arawak, Carib, Emerillon, Galibi (now called the Kalina), Palikour, Wayampi and Wayana. As of late 1990s there was evidence of an uncontacted group of Wayampi.

The dominant religion of French Guiana is Roman Catholicism; the Maroons and some Amerindian people maintain their own religions. The Hmong people are also mainly Catholic owing to the influence of missionaries who helped bring them to French Guiana.

Languages

The official language of French Guiana is French, but a number of other local languages exist. Regional languages include French Guiana creole, six Amerindian languages (Arawak, Palikur, Kali'na, Wayana, Wayampi, Emerillon), four Maroon dialects (Saramaka, Paramaccan, Boni, Djuka), as well as Hmong Njua. Other languages spoken include Portuguese, Hakka, Haitian Creole, Spanish and English.

Politics

French Guiana, as part of France, is part of the European Union, the largest landmass for an area outside of Europe (since Greenland left the European Community in 1985), with one of the longest EU external boundaries. Along with the Spanish enclaves in Africa of Ceuta and Melilla, it is one of only three European Union territories outside Europe that is not an island. As an integral part of France, its head of state is the President of the French Republic, and its head of Government is the Prime Minister of France. The French Government and its agencies have responsibility for a wide range of issues that are reserved to the National Executive, such as defense and external relations.

The President of France appoints a Prefect (resident at the Prefecture building in Cayenne) as his representative to head the local government of French Guiana. There are two legislative bodies: the 19-member General Council and the 34-member Regional Council, both elected.

French Guiana sends two deputies to the French National Assembly, one representing the commune (municipality) of Cayenne and the commune of Macouria, and the other representing the rest of French Guiana. This latter constituency is the largest in the French Republic by land area. French Guiana also sends one senator to the French Senate.

French Guiana has traditionally been conservative, though the socialist party has been increasingly successful in recent years.

A chronic issue affecting French Guiana is the influx of illegal immigrants and clandestine gold prospectors from Brazil and Suriname. The border between the department and Suriname is formed by the Maroni River, which flows through rain forest and is difficult for the Gendarmerie and the French Foreign Legion to patrol. There have been several phases launched by the French government to combat illegal gold mining in French Guiana, beginning with Operation Anaconda beginning in 2003, followed by Operation Harpie in 2008, 2009 and Operation Harpie Reinforce in 2010. Colonel Francois Muller, the commander of French Guiana's gendarme believes these operations have been successful. However, after each operation ends, Brazilian miners, Garimpeiros, return. Soon after Operation Harpie Reinforce began, an altercation took place between French authorities and Brazilian miners. On March 12th, 2010 a team of French soldiers and border police were attacked while returning from a successful operation, during which "the soldiers had arrested 15 miners, confiscated three boats, and seized 617 grams of gold... currently worth about $22,317." Garimpeiros returned to retrieve the lost loot and colleagues. "The soldiers fired warning shots and rubber "flash balls" but the miners managed to retake one of their boats and about 500 grams of gold. "The violent reaction by the garimpeiros can be explained by the exceptional take of 617 grams of gold, about 20 percent of the quantity seized in 2009 during the battle against illegal mining," said Phillipe Duporge, the director of French Guiana's border police, at a press conference the next day."

Transport

French Guiana's main international airport is Cayenne-Rochambeau Airport, located in the commune of Matoury, a southern suburb of Cayenne. There are three flights a day to Paris (Orly Airport), served by Air France, Air Caraïbes and CorsairFly. The flight time from Cayenne to Paris is 8 hours and 25 minutes, and from Paris to Cayenne it is 9 hours and 10 minutes. There are also flights to Fort-de-France, Pointe-a-Pitre, Port-au-Prince, Miami and Belem.

French Guiana's main seaport is the port of Degrad des Cannes, located on the estuary of the Mahury River, in the commune of Remire-Montjoly, a south-eastern suburb of Cayenne. Almost all of French Guiana's imports and exports pass through the port of Degrad des Cannes. Built in 1969, it replaced the old harbour of Cayenne which was congested and couldn't cope with modern traffic.

An asphalted road from Regina to Saint-Georges de l'Oyapock (a town by the Brazilian border) was opened in 2004, completing the road from Cayenne to the Brazilian border. It is now possible to drive on a fully paved road from Saint-Laurent-du-Maroni on the Surinamese border to Saint-Georges de l'Oyapock on the Brazilian border.

Following an international treaty between France and Brazil signed in July 2005, the Oyapock River Bridge over the Oyapock River (marking the border with Brazil) is currently being built and is due to open in 2010. This bridge will be the first land crossing ever opened between France and Brazil, and indeed between French Guiana and the rest of the world (there exists no other bridge crossing the Oyapock River, and no bridge crossing the Maroni River marking the border with Suriname - there is a ferry crossing to Albina, Suriname.). When the bridge is opened, it will be possible to drive uninterrupted from Cayenne to Macapa, the capital of the state of Amapa in Brazil.

3

Venezuela

Venezuela, officially called the Bolivarian Republic of Venezuela (Spanish: *Republica Bolivariana de Venezuela*), is a tropical country on the northern coast of South America. It is a continental mainland with numerous islands located off its coastline in the Caribbean Sea. The republic is a former Spanish colony, that won its independence in 1821.

Venezuela borders Guyana to the east, Brazil to the south, and Colombia to the west. Trinidad and Tobago, Grenada, St. Lucia, Barbados, Curacao, Bonaire, Aruba, Saint Vincent and the Grenadines and the Leeward Antilles lie just north, off the Venezuelan coast. Its size is 916,445 km² with an estimated population of 26,414,816. Its capital is Caracas. The colours of the Venezuelan flag are yellow, blue and red, in that order: the yellow stands for land wealth, the blue for the sea and sky of the country, and the red for the blood shed by the heroes of independence.

Venezuela has territorial disputes with Guyana (formerly United Kingdom), largely concerning the Essequibo area, and with Colombia concerning the Gulf of Venezuela. In 1895, after years of diplomatic attemps to solve the border dispute, from Venezuela, the dispute over the Essequibo River border flared up, it was submitted to a "neutral" commission (composed of United Kingdom, United States and Russian representatives and without a direct Venezuelan representative), which in 1899 decided mostly against Venezuela's claim. Venezuela is known widely for its petroleum industry, the environmental diversity of its territory, and its natural features. Venezuela is considered to be among the world's 17 most biodiverse countries, featuring diverse wildlife in a variety of protected habitats.

Venezuela is among the most urbanized countries in Latin America; the vast majority of Venezuelans live in the cities of the north, especially in the capital Caracas which is also the largest city. Other major cities include Maracaibo, Valencia, Maracay, Barquisimeto, Merida, Barcelona-Puerto La Cruz and Ciudad Guayana.

Etymology

In 1498, within the framework of his third voyage, Christopher Columbus sailed near the Orinoco Delta, then interned in the Gulf of Paria. Amazed, Columbus expressed in his moving letter to the Catholic Monarchs that he had reached the heaven on earth (the paradise), and confused by the unusual saltiness of the water, he wrote:

Great signs are these of the Terrestrial Paradise, for the site conforms to the opinion of the holy and wise theologians whom I have mentioned. And likewise, the [other] signs conform very well, for I have never read or heard of such a large quantity of fresh water being inside and in such close proximity to salt water; the very mild temperateness also corroborates this; and if the water of which I speak does not proceed from Paradise then it is an even greater marvel, because I do not believe such a large and deep river has ever been known to exist in this world.

His certainty of having attained Paradise made him name this region *Land of Grace*, a phrase which has become the country's nickname.

Nevertheless, the following year of 1499, an expedition led by Alonso de Ojeda visited the Venezuelan coast. The stilt houses in the area of Lake Maracaibo reminded the navigator Amerigo Vespucci of the city of Venice, (Italian: *Venezia*), so he named the region "*Venezuela*," meaning "little Venice" in Italian. The word has the same meaning in Spanish, where the suffix *-zuela* is used as a diminutive term (e.g., *plaza / plazuela*, *cazo / cazuela*); thus, the term's original sense would have been that of a "little Venice."

Nonetheless, although the Vespucci story remains the most popular and accepted version of the origin of the country's name, a different reason for the name comes up in the account of Martin Fernandez de Enciso, a member of the Vespucci and Ojeda crew. In his work *Summa de Geografia,* he states that they found an indigenous population who called themselves the "*Veneciuela,*" which suggests that the name "Venezuela" may have evolved from the native word.

History

Human habitation of Venezuela could have commenced at least 15,000 years ago from which period leaf-shaped tools, together with chopping and plano-convex scraping implements, have been found exposed on the high riverine terraces of the Rio Pedregal in western Venezuela. Late Pleistocene hunting artifacts, including spear tips, have been found at a similar series of sites in northwestern Venezuela known as "El Jobo"; according to radiocarbon dating, these date from 13,000 to 7,000 BC.

Venezuela was first colonized by Spain in 1522 in what is now Cumana. These portions of eastern Venezuela were incorporated into New Andalusia. Administered by the Audiencia of Santo Domingo since the early 16th century, most of Venezuela became part of the Viceroyalty of New Granada in the early 18th century, and was then reorganized as an autonomous Captaincy General starting in 1776.

In the 16th century, during the Spanish colonization, indigenous peoples such as many of the Mariches, themselves descendants of the Caribs rejected paganism and embraced Roman Catholicism. Some Spaniards treated the natives harshly. Indian caciques (leaders) such as Guaicaipuro and Tamanaco attempted to resist Spanish incursions, but were ultimately defeated; Tamanaco was put to death by order of Caracas' founder Diego de Losada.

After a series of unsuccessful uprisings, Venezuela—under the leadership of Francisco de Miranda, a Venezuelan marshal who had fought in the American Revolution and the French Revolution—declared independence on 5 July 1811. This began the Venezuelan War of Independence. However, a devastating earthquake that struck Caracas in 1812, together with the rebellion of the Venezuelan *llaneros*, helped bring down the first Venezuelan republic. A second Venezuelan republic, proclaimed on 7 August 1813, lasted several months before being crushed as well.

Sovereignty was only attained after Simon Bolivar, aided by Jose Antonio Paez and Antonio Jose de Sucre, won the Battle of Carabobo on 24 June 1821. Jose Prudencio Padilla and Rafael Urdaneta's victory in the Battle of Lake Maracaibo on 24 July 1823, helped seal Venezuelan independence. New Granada's congress gave Bolivar control of the Granadian army; leading it, he liberated several countries and founded Gran Colombia.

Sucre, who won many battles for Bolivar, went on to liberate Ecuador and later become the second president of Bolivia. Venezuela remained part of Gran Colombia until 1830, when a rebellion led by Paez allowed the proclamation of a newly independent Venezuela; Paez became the first president of the new republic. Two decades of warfare had cost the lives of between a quarter and a third of the Venezuelan population, which in 1830 numbered no more than 800,000.

Much of Venezuela's nineteenth century history was characterized by political turmoil and dictatorial rule.

20th Century

During the first half of the 20th century, *caudillos* (military strongmen) continued to dominate, though they generally allowed for mild social reforms and promoted economic growth. Following the death of Juan Vicente Gómez in 1935 and the demise of *caudillismo* (authoritarian rule), pro-democracy movements eventually forced the military to withdraw from direct involvement in national politics in 1958. Since that year, Venezuela has had a series of democratically elected governments.

The discovery of massive oil deposits during World War I prompted an economic boom that lasted into the 1980s; by 1935, Venezuela's per capita gross domestic product was Latin America's highest. After World War II the globalization and heavy immigration from Southern Europe (mainly from Spain, Italy, Portugal) and poorer Latin American countries markedly diversified Venezuelan society.

The huge public spending and accumulation of internal and external debts during the Petrodollar years of the 1970s and early 1980s, followed by the collapse of oil prices during the 1980s, crippled the Venezuelan economy. As the government started to devaluate the currency in February 1983 in order to face its financial obligations, Venezuelans' real standard of living fell dramatically. A number of failed economic policies and increasing corruption in government led to rising poverty and crime, worsening social indicators, and increased political instability. Corruption remains a problem; Venezuela was ranked near the bottom of countries in the Corruptions Perceptions Index in 2009.

In February 1992 Hugo Chavez, an army paratrooper, staged a coup d'etat attempt seeking to overthrow the government of President Carlos Andres Perez. Chavez failed and was placed in jail. In November 1992, another unsuccessful coup attempt occurred, organized by groups loyal to Chavez remaining in the armed forces. Chavez was pardoned in March 1994 by president Rafael Caldera, with a clean slate and his political rights intact.

In 1998, Chavez was elected president. His reform program, which he later called the "Bolivarian Revolution", was aimed at redistributing the benefits of Venezuela's oil wealth to the lower socio-economic groups by using it to fund programs such as health care and education, but has encountered great criticism by the previous establishment. In April 2002 he suffered a *coup d'etat*. He was returned to power after two days as a result of popular demonstrations in his favour and actions by the military. Chavez has also survived an all-out national strike that lasted more than two months in December 2002 – February 2003, including a strike/lockout in the state oil company PDVSA, and a recall referendum in August 2004. He was elected for another term in December 2006.

Economy

Unemployment data for 2005 is the CIA World Factbook estimate. 1 trillion = 1,000,000,000,000. The vertical scales do not start at 0 to make more details visible. Oil production figures from IEA in millions of barrels per day.

The petroleum sector dominates Venezuela's mixed economy, accounting for roughly a third of GDP, around 80%

of exports and more than half of government revenues. Gold, diamonds and iron ore are mined as well. Venezuela contains some of the largest oil and natural gas reserves in the world. It consistently ranks among the top ten crude oil producers in the world.

The country's main petroleum deposits are located around and beneath Lake Maracaibo, the Gulf of Venezuela (both in Zulia), and in the Orinoco River basin (eastern Venezuela), where the country's largest reserve is located. Venezuela has the least expensive petrol in the world because of its high government subsidies.

Inflation has been a problem. It was expected to slow to 26% annually in 2009, according to the president of the national bank, Nelson Merentes.

Personal Income

Per capita GDP for 2008 was US$13,500, ranking it 84th in the world. About 30 % of the population of the country live on less than US$ 2 per day.

Petroleum and other Resources

When oil was discovered at the Maracaibo strike in 1922, Venezuela's dictator Juan Vicente Gómez allowed Americans to write Venezuela's petroleum law. But oil history was made in 1943 when Standard Oil of New Jersey accepted a new agreement in Venezuela based on the 50–50 principle, "a landmark event."

Terms even more favourable to Venezuela were negotiated in 1945, after a coup brought to power a left-leaning government that included Juan Pablo Perez Alfonso. In 1958 a new government again included Perez Alfonso, who devised a plan for the international oil cartel that would become OPEC. In 1973 Venezuela voted to nationalize its oil industry outright, effective 1 January 1976, with Petroleos de Venezuela (PDVSA) taking over and presiding over a number of holding companies; in subsequent years, Venezuela built a vast refining and marketing system in the U.S. and Europe. Economic prospects remain highly dependent on oil prices and the export of petroleum. A founding member of the Organization of Petroleum Exporting Countries (OPEC), Venezuela reasserted its leadership within the organization during its year as OPEC's president, hosting the organization's Second Leadership Conference in 40 years, as well as having its former Minister of Energy, Alvaro Silva Calderon, appointed as Secretary General.

The collapse of oil prices in 1997–98 prompted the Rodriguez administration to expand OPEC-inspired production cuts in an effort to raise world oil prices. In 2002, this sector accounted for roughly a quarter of GDP, 73% of export earnings, and about half of central government's operating revenues. Venezuela is the fourth-leading supplier of imported crude and refined petroleum products to the United States.

The Government of Venezuela has opened up much of the hydrocarbon sector to foreign investment, promoting multi-billion dollar investment in heavy oil production, reactivation of old fields, and investment in several petrochemical joint ventures. Almost 60 foreign companies representing 14 different countries participate in one or more aspects of Venezuela's oil sector.

The Venezuelan national oil company Petroleos de Venezuela, S.A. (PDVSA) and foreign oil companies have signed 33 operating contracts for marginal fields in three bidding rounds. New legislation dealing with natural gas and petrochemicals is further opening the sector. A new domestic retail competition law, however, disappointed investors who had been promised market-determined prices.

On 13 November 2001, under the enabling law authorized by the National Assembly, President Chavez enacted the new Hydrocarbons Law, which came into effect in January 2002. This law replaced the Hydrocarbons Law of 1943 and the Nationalization Law of 1975. Among other things, the new law provided that all oil production and distribution activities were to be the domain of the Venezuelan state, with the exception of joint ventures targeting extra-heavy crude oil production. Under the new Hydrocarbons Law, private investors can own up to 49% of the capital stock in joint ventures involved in upstream activities. The new law also provides that private investors may own up to 100% of the capital stock in ventures concerning downstream activities, in addition to the 100%

already allowed for private investors with respect to gas production ventures, as previously promulgated by the National Assembly.

During the December 2002-February 2003 all-out national strike where managers and skilled highly paid technicians of PDVSA shut down the plants and left their posts, petroleum production and refining by PDVSA almost ceased. At the same time, many business owners across Venezuela closed down their stores, both actions aimed at ousting Chavez from government. After more than 60 days of getting nowhere the strike died off, and activities eventually were slowly restarted by returning and substitute oil workers. Out of a total of 45,000 PDVSA management and workers, some 19,000 were subsequently dismissed with no compensation; many of whom were managers and highly paid professionals and technicians who thereafter were banned from working in the petroleum industry, even indirectly.

Manufacturing, Agriculture, and Trade

Manufacturing contributed 17% of GDP in 2006. Venezuela manufactures and exports steel, electronics, aluminum, automobiles, textiles, apparel, beverages, and foodstuffs. It produces cement, tires, paper, fertilizer, and assembles cars both for domestic and export markets.

Agriculture accounts for approximately 3% of GDP, 10% of the labour force, and at least one-fourth of Venezuela's land area. Venezuela exports rice, corn, fish, tropical fruit, coffee, beef, and pork. The country is not self-sufficient in most areas of agriculture; Venezuela imports about two-thirds of its food needs. In 2005 87, U.S. firms exported $347 million worth of agricultural products, including wheat, corn, soybeans, soybean meal, cotton, animal fats, vegetable oils, and other items to make Venezuela one of the top two U.S. markets in South America.

Due to petroleum exports, Venezuela usually posts a trade surplus. In recent years, nonpetroleum exports have been growing rapidly but still constitute only about one-fourth of total exports. The United States is Venezuela's leading trade partner although Brazil is expected to surpass the U.S. by 2011. During 2002, the United States exported $4.4 billion in goods to Venezuela, making it the 25th-largest market for the U.S. Including petroleum products, Venezuela exported $15.1 billion in goods to the U.S., making it its 14th-largest source of goods.

Geography

Venezuela is a country in South America, and part of Caribbean South America, bordering the Caribbean Sea and the North Atlantic Ocean, between Colombia and Guyana. It is situated on major sea and air routes linking North and South America. Located at the northernmost end of South America, Venezuela has a total area of 916,445 square kilometers and a land area of 882,050 square kilometers, about twice the size of California. Shaped roughly like an inverted triangle, the country has a 2,800-kilometer coastline and is bounded on the north by the Caribbean Sea and the Atlantic Ocean, on the east by Guyana, on the south by Brazil, and on the west by Colombia. Venezuela is 352,144 sq. mi.

Topography

Most observers describe Venezuela in terms of four fairly well-defined regions: the Maracaibo lowlands in the northwest, the northern mountains extending in a broad east-west arc from the Colombian border along the Caribbean Sea, the wide Orinoco plains (llanos) in central Venezuela, and rank highly dissected Guiana highlands in the southeast.

The Maracaibo lowlands form a large spoon-shaped oval bounded by mountains on three sides and open to the Caribbean on the north. The area is remarkably flat with only a gentle slope toward the center and away from the mountains that border the region. Lake Maracaibo occupies much of the lower-lying territory. Areas around the southern part of Lake Maracaibo are swampy, and, despite the rich agricultural land and significant petroleum deposits, the area was still thinly populated in 1990.

Climate

Although the country lies wholly within the tropics, its climate varies from tropical humid to alpine, depending on the elevation, topography, and the direction and intensity of prevailing winds. Seasonal variations are marked less by temperature than by rainfall. Most of the country has a distinct rainy season; the rainy period (May through November) is commonly referred to as winter and the remainder of the year as summer.

The country falls into four horizontal temperature zones based primarily on elevation. In the tropical zone—below convert | 800 meters—temperatures are hot, with yearly averages ranging between 26 and 28 °C (78.8 and 82.4 °F). The temperate zone ranges between 800 and 2,000 meters (2,625 and 6,562 ft) with averages from convert | 12 | to | 25 | °C | °F | 1; many of Venezuela's cities, including the capital, lie in this region. Colder conditions with temperatures from 9 to 11 °C (48.2 to 51.8 °F) are found in the cool zone between 2,000 and 3,000 meters (6,562 and 9,843 ft). Pastureland and permanent snowfield with yearly averages below 8 °C (46 °F) cover land above 3,000 meters (9,843 ft) in the high mountain areas known as the paramos.

Average yearly rainfall amounts in the lowlands and plains range from a semiarid 430 millimeters (16.9 in) in the western part of the Caribbean coastal areas to around 1,000 millimeters (39.4 in) in the Orinoco Delta. Rainfall in mountainous areas varies considerably; sheltered valleys receive little rain, but slopes exposed to the northeast trade winds experience heavy rainfall. Caracas averages 750 millimeters (29.5 in) of precipitation annually, more than half of it falling from June through August.

Polarography

The Orinoco is by far the most important of the more than 1,000 rivers in the country. Flowing more than 2,500 kilometers to the Atlantic from its source in the Guiana highlands at the Brazilian border, the Orinoco is the world's eighth largest river and the largest in South America after the Amazon. Its flow varies substantially by season, with the high water level in August exceeding by as much as thirteen meters the low levels of March and April. During low water periods, the river experiences high and low tides for more than 100 kilometers upstream from Ciudad Guayana. For most of the river's course, the gradient is slight. Downstream from its headwaters, it splits into two; one-third of its flow passes through the Brazo Casiquiare (Casiquiare Channel) into a tributary of the Amazon, and the remainder passes into the main Orinoco channel. This passageway allows vessels with shallow drafts to navigate from the lower Orinoco to the Amazon River system after unloading and reloading on either side of two falls on the Orinoco along the Colombian border.

Most of the rivers rising in the northern mountains are flowing southeastward to the Rio Apure, a tributary of the Orinoco. From its headwater, the Apure crosses the llanos in a generally eastward direction. Few rivers flow into it from the poorly drained region south of the river and much of this area near the Colombian border is swampland.

The other major Venezuelan river is the fast-flowing Caroni, which originates in the Guiana highlands and flows northward into the Orinoco upstream from Ciudad Guyana. The Caroni is capable of producing as much hydroelectric power as any river in Latin America and has contributed significantly to the nation's electric power production. Electricity generated by the Caroni was one of the factors encouraging industrialization of the northern part of the Guiana highlands and the lower Orinoco valley.

Lake Maracaibo, the largest lake in Latin America, occupies the central 13,500 square kilometers of the Maracaibo lowlands. The low swampy shores of the lake and areas beneath the lake itself hold most of Venezuela's rich petroleum deposits. The lake is shallow, with an average depth of ten meters, and separated from the Caribbean by a series of islands and sandbars. In 1955 a 7.5-meter channel was cut through the sandbars to facilitate shipping between the lake and the Caribbean. The channel also allows salt water to mix with the yellowish fresh water of the lake, making the northern parts brackish and unsuited for drinking or irrigation.

Venezuela is a country in the north of South America, bordering the Caribbean Sea. It is bounded on the south by Brazil, and on the west by Colombia. Venezuela has a total area of 916,445 square kilometres (353,841 sq mi) and a land area of 882,050 square kilometres (340,560 sq mi), about twice the size of California. Shaped roughly like an inverted triangle, the country has a 2,800 kilometres (1,700 mi) coastline. With 2,800 kilometres (1,740 mi) of coastline, there is a variety of landscapes. The extreme northeastern extensions of the Andes reach into Venezuela's northwest and continue along the northern Caribbean coast. Pico Bolivar, the nation's highest point at 4,979 metres (16,335 ft), lies in this region. The country's center is characterized by the *llanos*, which are extensive plains that stretch from the Colombian border in the far west to the Orinoco River delta in the east.

To the south, the dissected Guiana Highlands contains the northern fringes of the Amazon Basin and Angel Falls, the world's highest waterfall as well as tepuis, large table-like mountains. The Orinoco, with its rich alluvial soils, binds the largest and most important river system of the country; it originates in one of the largest watersheds in Latin America. The Caroni and the Apure are other major rivers.

The Insular Region includes all of Venezuela's island possessions: Nueva Esparta and the various Federal Dependencies. The Deltaic System, which forms a triangle covering Delta Amacuro, projects northeast into the Atlantic Ocean.

The country can be further divided into ten geographical areas, some corresponding to climatic and biogeographical regions. In the north are the Venezuelan Andes and the Coro region, a mountainous tract in the northwest, holds several sierras and valleys. East of it are lowlands abutting Lake Maracaibo and the Gulf of Venezuela. The Central Range runs parallel to the coast and includes the hills surrounding Caracas; the Eastern Range, separated from the Central Range by the Gulf of Cariaco, covers all of Sucre and northern Monagas.

Geology

Venezuela's mainland rests on the South American Plate.

Flora and Fauna

Venezuela lies within the Neotropic ecozone; large portions of the country were originally covered by moist broadleaf forests. One of seventeen megadiverse countries and among the top twenty countries in terms of endemism, some 38% of the over 21,000 plant species are unique to the country; 23% of reptilian and 50% of amphibian species are also endemic. Venezuela hosts significant biodiversity across habitats ranging from xeric scrublands in the extreme northwest to coastal mangrove forests in the northeast. Its cloud forests and lowland rainforests are particularly rich, for example hosting over 25,000 species of orchids. These include the *flor de mayo* orchid (*Cattleya mossiae*), the national flower.

Venezuela's national tree is the *araguaney*, whose characteristic lushness after the rainy season led novelist Romulo Gallegos to name it *«[l]a primavera de oro de los araguaneyes»* ("the golden spring of the *araguaneyes*"). Notable mammals include the giant anteater, jaguar, and the capybara, the world's largest rodent. More than half of Venezuelan avian and mammalian species are found in the Amazonian forests south of the Orinoco.

Manatees, Boto river dolphins, and Orinoco crocodiles, which have been reported to reach up to 6.6 metres (22 ft) in length, are notable aquatic species. Venezuela hosts a total of 1,417 bird species, 48 of which are endemic. Important birds include ibises, ospreys, kingfishers, and the yellow-orange turpial, the national bird.

In recent decades, logging, mining, shifting cultivation, development, and other human activities have posed a major threat to Venezuela's wildlife; between 1990 and 2000, 0.40% of forest cover was cleared annually. In response, federal protections for critical habitat were implemented; for example, 20% to 33% of forested land is protected. The country has a biosphere reserve that is part of the World Network of Biosphere Reserves; five wetlands are registered under the Ramsar Convention. In 2003, 70% of the nation's land was under conservation management in over 200 protected areas, including 43 national parks.

Climate

Though Venezuela is entirely situated in the tropics, its climate varies from humid low-elevation plains, where average annual temperatures range as high as 28 °C (82.4 °F), to glaciers and highlands (the *paramos*) with an average yearly temperature of 8 °C (46.4 °F). Annual rainfall varies between 430 millimetres (16.9 in) in the semiarid portions of the northwest to 1,000 millimetres (39.4 in) in the Orinoco Delta of the far east. Most precipitation falls between June and October (the rainy season or "winter"); the drier and hotter remainder of the year is known as "summer", though temperature variation throughout the year is not as pronounced as at temperate latitudes.

The country falls into four horizontal temperature zones based primarily on elevation, having Tropical, Dry, Temperate with Dry Winters, and Polar (Alpine tundra) climates, amongst others. In the tropical zone—below 800 meters / 2,625 feet—temperatures are hot, with yearly averages ranging between 26 and 28 °C (78.8 and 82.4 °F). The temperate zone ranges between 800 and 2,000 meters (2,625 and 6,562 ft) with averages from 12 to 25 °C (53.6 to 77 °F); many of Venezuela's cities, including the capital, lie in this region. Colder conditions with temperatures from 9 to 11 °C (48.2 to 51.8 °F) are found in the cool zone between 2,000 and 3,000 meters (6,562 and 9,843 ft), especially in the venezuelan Andes, where Pastureland and permanent snowfield with yearly averages below 8 °C (46 °F) cover land above 3,000 meters (9,843 ft) in the high mountain areas known as the paramos.

Subdivisions

Venezuela is divided into 23 states (*estados*), a capital district (*distrito capital*) corresponding to the city of Caracas, and the Federal Dependencies (*Dependencias Federales*, a special territory). Venezuela is further subdivided into 335 municipalities (*municipios*); these are subdivided into over one thousand parishes (*parroquias*). The states are grouped into nine administrative regions (*regiones administrativas*), which were established in 1969 by presidential decree; in addition, Venezuela has historically claimed and continues to claim all Guyanese territory west of the Essequibo River; this 159,500 square kilometres (61,583 sq mi) tract was dubbed *Guayana Esequiba* or the *Zona en Reclamation* (the "zone to be reclaimed").

Demographics

Venezuela's birth rate is among the highest in South America, after Bolivia, Paraguay and French Guyana.

Since 1930, Venezuelan census does not contain information about ethnicity, so only rough estimates are available. Some 60% of the population are Mestizo, defined as a mixture of European and Amerindian; approximately 30% are European, mostly of Spanish, Italian, Portuguese, Lebanese, and German descent.

Two of the main Amerindian tribes located in the country are the Wayuu, located in the west, in Zulia, and the Timotocuicas, also in the west, in Merida, in the Andes. Other important groups include Afro-Venezuelans, though their numbers are unclear due to poor census data.

Asians make up a small percentage of the population. About 1% of Venezuelans are indigenous. These groups were joined by sponsored migrants from throughout Europe and neighboring parts of South America by the mid-20th century economic boom.

According to the *World Refugee Survey 2008*, published by the U.S. Committee for Refugees and Immigrants, Venezuela hosted a population of refugee and asylum seekers from Colombia numbering 252,200 in 2007. 10,600 new asylum seekers entered Venezuela in 2007. Between 500,000 and one million illegal immigrants are estimated to be living in the Venezuela.

About 85% of the population live in urban areas in northern Venezuela; 73% live less than 100 kilometres (62 mi) from the coastline. Though almost half of Venezuela's land area lies south of the Orinoco, only 5% of Venezuelans live there. The largest and most important city south of the Orinoco is Ciudad Guayana, which is the sixth most populous conurbation.

The national and official language is Spanish; 31 indigenous languages are also spoken, including Guajibo, Pemon, Warao, Wayuu, and the various Yanomaman languages.

Religion

According to government estimates, 92% of the population is at least nominally Roman Catholic, and the remaining 8% are Protestant, or a member of another religion. The Venezuelan Evangelical Council estimates that Evangelical Protestants constitute 10% of the population.

Government

Executive

The Venezuelan president is elected by a vote with direct and universal suffrage, and is both head of state and head of government. The term of office is six years, and (as of 15 February 2009) a president may be re-elected an unlimited number of times. The president appoints the vice-president and decides the size and composition of the Cabinet and makes appointments to it with the involvement of the legislature. The president can ask the legislature to reconsider portions of laws he finds objectionable, but a simple parliamentary majority can override these objections.

Legislative

The unicameral Venezuelan parliament is the *Asamblea Nacional* ("National Assembly"). Its 167 deputies, of which three are reserved for indigenous people, serve five-year terms and may be re-elected for a maximum of two additional terms. They are elected by popular vote through a combination of party lists and single member constituencies.

The voting age in Venezuela is 18 and older. Voting is not compulsory.

Judicial

The highest judicial body is the Supreme Tribunal of Justice or *Tribunal Supremo de Justicia*, whose magistrates are elected by parliament for a single twelve-year term. The National Electoral Council (*Consejo Nacional Electoral*, or *CNE*) is in charge of electoral processes; it is formed by five main directors elected by the National Assembly. Supreme Court president Luisa Estela Morales said in December 2009 that Venezuela had moved away from "a rigid division of powers" toward a system characterized by "intense coordination" between the branches of government. Morales clarified that each power must be independent adding that "one thing is separation of powers and another one is division".

Politics

There are currently two major blocs of political parties in Venezuela: the incumbent leftist bloc United Socialist Party of

Venezuela (PSUV), its major allies Fatherland for All (PPT) and the Communist Party of Venezuela (PCV), and the opposition bloc led by A New Era (UNT) together with its allied parties Project Venezuela, Justice First, Movement for Socialism and others.

Following the fall of Marcos Perez Jiménez in 1958, Venezuelan politics was dominated by the third-way Christian democratic COPEI and the center-left social democratic Democratic Action (AD) parties; this two-party system was formalized by the *puntofijismo* arrangement. However, this system has been sidelined following the 1998 election of current President Hugo Chavez, which started what he calls the Bolivarian Revolution.

Most of the political opposition boycotted the 2005 parliamentary election. Consequently, Hugo Chavez's MVR-led bloc secured all 167 seats in the National Assembly. Then the MVR voted to dissolve itself and join the new United Socialist Party of Venezuela, while Chavez requested that MVR-allied parties merge themselves into it as well. The National Assembly has twice voted to grant Chavez the ability to rule by decree for several months in several broadly defined areas, once in 2000 and again in 2007. This power had only rarely been granted to previous administrations, and then only for extraordinary circumstances and for a short time. Chavez has established alliances with several Latin American countries which have elected leftist governments, such as Argentina, Bolivia, Brazil, Ecuador, Guatemala, Nicaragua, El Salvador, Honduras and Paraguay.

There is a controversy regarding the independence of justice, the opposition arguing that the justice system is used as a weapon against opponents. The government denies that.

Health

Venezuela has a national universal health care system that is free of charge. The current government has created a program to expand access to health care known as Mission Barrio Adentro.

Infant mortality in Venezuela stood at 16 deaths per 1,000 births in 2004, lower than the South American average (by comparison, the U.S. stands at 5 deaths per 1,000 births in

2006). Child malnutrition (defined as stunting or wasting in children under age five) stands at 17%; Delta Amacuro and Amazonas have the nation's highest rates. According to the United Nations, 32% of Venezuelans lack adequate sanitation, primarily those living in rural areas. Diseases ranging from typhoid, yellow fever, cholera, hepatitis A, hepatitis B, and hepatitis D are present in the country.

In 2009, Chavez requested that the people take three minute showers because of a water shortage. Other sources point to underinvestment as an added problem.

Venezuela has a total of 150 plants for sewage treatment. However still 13% of the population lack access to drinking water but this number seems to be dropping.

Travelers to Venezuela are advised to obtain vaccinations for a variety of diseases including typhoid, yellow fever, cholera, hepatitis A, hepatitis B and hepatitis D. In a cholera epidemic of contemporary times in the Orinoco Delta, Venezuela's political leaders were accused of racial profiling of their own indigenous people to deflect blame from the country's institutions, thereby aggravating the epidemic.

The nation has one of the worst murder rates in the world, 31.6 homicides per 100,000 inhabitants. The capital Caracas has the second greatest homicide rate of any large city in the world, with 92 homicides per 100,000 residents. In 2008, the Interior Minister resigned, apparently because he had failed to change voter perceptions that crime was out of control. In 2008, polls indicated that crime was the number one concern of voters.

Foreign Relations

Throughout most of the 20th century, Venezuela maintained friendly relations with most Latin American and Western nations. Relations between Venezuela and the United States government worsened in 2002, after the 2002 Venezuelan coup d'etat attempt during which the U.S. government recognized the short-lived interim presidency of Pedro Carmona. Correspondingly, ties to various Latin American and Middle Eastern countries not allied to the U.S. have strengthened. Venezuela seeks alternative hemispheric integration via such proposals as the Bolivarian Alternative for the Americas trade proposal and the newly launched pan-Latin American television network tele SUR. Venezuela is one of the four nations in the world — along with Russia, Nicaragua and Nauru — to have recognized the independence of Abkhazia and South Ossetia. Venezuela was a proponent of OAS's decision to adopt its Anti-Corruption Convention, and is actively working in the Mercosur trade bloc to push increased trade and energy integration. Globally, it seeks a "multi-polar" world based on strengthened ties among Third World countries.

Military

The National Armed Forces of the Bolivarian Republic of Venezuela (Fuerza Armada Nacional, FAN) are the overall unified military forces of Venezuela. It includes over 129,150 men and women, under Article 328 of the Constitution, in 5 components of Ground, Sea and Air. The components of the National Armed Forces are: the Venezuelan Army, the Venezuelan Navy, the Venezuelan Air Force, the Venezuelan National Guard, and the Venezuelan National Militia

As of 2008, a further 600,000 soldiers were incorporated into a new branch, known as the Armed Reserve. The President of Venezuela is the commander-in-chief of the national armed forces. The main roles of the armed forces are to defend the sovereign national territory of Venezuela, airspace, and islands, fight against drug trafficking, to search and rescue and, in the case of a natural disaster, civil protection. All men that are citizens of Venezuela have a constitutional duty to register for the military at the age of 18, which is the age of majority in Venezuela.

Culture

Venezuela's heritage, art, and culture have been heavily influenced by the Caribbean context. These elements extend to its historic buildings, architecture, art, landscape, boundaries, and monuments. Venezuelan culture has been shaped by indigenous, Spanish and African influences. Before this period, indigenous culture was expressed in art (petroglyphs), crafts, architecture (*shabonos*), and social organization. Aboriginal culture was subsequently assimilated by Spaniards; over the years, the hybrid culture had diversified by region.

Venezuelan art was initially dominated by religious motifs but began emphasizing historical and heroic representations in the late 19th century, a move led by Martin Tovar y Tovar. Modernism took over in the 20th century. Notable Venezuelan artists include Arturo Michelena, Cristobal Rojas, Armando Reveron, Manuel Cabre; the kinetic artists Jesus-Rafael Soto and Carlos Cruz-Diez; and contemporary artist Yucef Merhi.

Venezuelan literature originated soon after the Spanish conquest of the mostly pre-literate indigenous societies; it was dominated by Spanish influences. Following the rise of political literature during the War of Independence, Venezuelan Romanticism, notably expounded by Juan Vicente Gonzalez, emerged as the first important genre in the region. Although mainly focused on narrative writing, Venezuelan literature was advanced by poets such as Andres Eloy Blanco and Fermin Toro.

Major writers and novelists include Romulo Gallegos, Teresa de la Parra, Arturo Uslar Pietri, Adriano Gonzalez Leon, Miguel Otero Silva, and Mariano Picon Salas. The great poet and humanist Andres Bello was also an educator and intellectual. Others, such as Laureano Vallenilla Lanz and Jose Gil Fortoul, contributed to Venezuelan Positivism.

Carlos Raul Villanueva was the most important Venezuelan architect of the modern era; he designed the Central University of Venezuela, (a World Heritage Site) and its Aula Magna. Other notable architectural works include the Capitolio, the Baralt Theatre, the Teresa Carreno Cultural Complex, and the General Rafael Urdaneta Bridge.

Indigenous musical styles of Venezuela are exemplified by the groups Un Solo Pueblo and Serenata Guayanesa. The national musical instrument is the cuatro. Typical musical styles and pieces mainly emerged in and around the *llanos* region, including *Alma Llanera* (by Pedro Elias Gutierrez and Rafael Bolivar Coronado), *Florentino y el Diablo* (by Alberto Arvelo Torrealba), *Concierto en la Llanura* by Juan Vicente Torrealba, and *Caballo Viejo* (by Simon Díaz).

The Zulian *gaita* is also a popular style, generally performed during Christmas. The national dance is the *joropo*. Teresa Carreno was a world-famous 19th century piano virtuosa. In the last years, Classical Music has had great performances. The Simon Bolivar Youth Orchestra has realized excellent presentations in many European concert halls, notably at the 2007 Proms, and has received honors of the public.

Baseball is Venezuela's most popular sport, although football (soccer), spearheaded by the Venezuela national football team, is gaining influence.

Venezuela is well-known for its successions in beauty pageants. Miss Venezuela is a big event in the country, and Venezuela has received 5 Miss World, 6 Miss Universe and 5 Miss International titles.

The World Values Survey has consistently shown Venezuelans to be among the happiest people in the world, with 55% of those questioned saying they were "very happy".

Education

In 2008, 95.2% of the adult population was literate.. Net primary school enrollment rate was at 91 % in 2005. Net secondary enrollment rate was at 63 % in 2005. Venezuela has a number of universities.

4

Colombia

Colombia, officially the Republic of Colombia (Spanish: *Republica de Colombia*, pronounced [reepualika oe koelombja] (listen)), is a constitutional republic in northwestern South America. Colombia is bordered to the east by Venezuela and Brazil; to the south by Ecuador and Peru; to the north by the Caribbean Sea; to the northwest by Panama; and to the west by the Pacific Ocean. Colombia also shares maritime borders with Venezuela, Jamaica, Haiti, the Dominican Republic, Honduras, Nicaragua and Costa Rica. With a population of over 45 million people, Colombia has the 29th largest population in the world and the second largest in South America, after Brazil. Colombia has the third largest Spanish-speaking population in the world after Mexico and Spain.

The territory of what is now Colombia was originally inhabited by indigenous nations including the Muisca, Quimbaya, and Tairona. The Spanish arrived in 1499 and initiated a period of conquest and colonization killing or taking as slaves almost 90% of the native population, and then creating the Viceroyalty of New Granada (comprising modern-day Colombia, Venezuela, Ecuador, the northwest region of Brazil and Panama) with its capital in Bogota. Independence from Spain was won in 1819, but by 1830 "Gran Colombia" had collapsed with the secession of Venezuela and Ecuador. What is now Colombia and Panama emerged as the Republic of New Granada. The new nation experimented with federalism as the Granadine Confederation (1858), and then the United States of Colombia (1863), before the Republic of Colombia was finally declared in 1886. Panama seceded in 1903 under pressure to fulfill financial responsibilities towards the United States government to build the Panama Canal.

Colombia has a long tradition of constitutional government. The Liberal and Conservative parties, founded in 1848 and 1849 respectively, are two of the oldest surviving political parties in the Americas. However, tensions between the two have frequently erupted into violence, most notably in the Thousand Days War (1899–1902) and La Violencia, beginning in 1948. Since the 1960s, government forces, left-wing insurgents and right-wing paramilitaries have been engaged in the continent's longest-running armed conflict. Fuelled by the cocaine trade, this escalated dramatically in the 1980s. However, in the recent decade (2000s) the violence has decreased significantly. Many paramilitary groups have demobilized as part of a controversial peace process with the government, and the guerrillas have lost control in many areas where they once dominated. Meanwhile Colombia's homicide rate, for many years one of the highest in the world, has almost halved since 2002. Assassinations of trade union members have also been significantly reduced since the 1990s but unionists continue to be threatened and murdered.

Colombia is a standing middle power with the fourth largest economy in Latin America, although income inequality is prevalent and wealth is poorly redistributed.

Colombia is very ethnically diverse, and the interaction between descendants of the original native inhabitants, Spanish colonists, Africans brought as slaves and twentieth-century immigrants from Europe and the Middle East has produced a rich cultural heritage. This has also been influenced by Colombia's varied geography. The majority of the urban centres are located in the highlands of the Andes mountains, but Colombian territory also encompasses Amazon rainforest, tropical grassland and both Caribbean and Pacific coastlines. Ecologically, Colombia is one of the world's 17 megadiverse countries (the most biodiverse per unit area).

Etymology

The word "Colombia" comes from Christopher Columbus (Spanish: *Cristobal Colon*). It was conceived by the Venezuelan revolutionary Francisco de Miranda as a reference to all the New World, but especially to those territories and colonies under Spanish and Portuguese rule. The name was later adopted by the Republic of Colombia of 1819, formed out of the territories of the old Viceroyalty of New Granada (modern-day Colombia, Panama, Venezuela and Ecuador).

In 1835, when Venezuela and Ecuador broke away, the Cundinamarca region that remained became a new country — the Republic of New Granada. In 1858 New Granada officially changed its name to the Grenadine Confederation, then in 1863 the United States of Colombia, before finally adopting its present name — the Republic of Colombia — in 1886.

Geography

Colombia is bordered to the east by Venezuela and Brazil; to the south by Ecuador and Peru; to the north by Panama and the Caribbean Sea; and to the west by Ecuador and the Pacific Ocean.

Part of the Pacific Ring of Fire, a region of the world subject to earthquakes and volcanic eruptions, Colombia is dominated by the Andes mountains. Beyond the Colombian Massif (in the south-western departments of Cauca and Marino) these are divided into three branches known as *cordilleras* (from the Spanish for "rope"): the *Cordillera Occidental*, running adjacent to the Pacific coast and including the city of Cali; the *Cordillera Central*, running between the Cauca and Magdalena river valleys (to the west and east respectively) and including the cities of Medullin, Manizales, Pereira and Armenia, Quindio ; and the *Cordillera Oriental*, extending north east to the Guajira Peninsula and including Bogota, Bucaramanga and Cucuta. Peaks in the *Cordillera Occidental* exceed 13,000 ft (3,962 m), and in the *Cordillera Central* and *Cordillera Oriental* they reach 18,000 ft (5,486 m). At 8,500 ft (2,591 m), Bogota is the highest city of its size in the world.

East of the Andes lies the savanna of the *Llanos*, part of the Orinoco River basin, and, in the far south east, the jungle of the Amazon rainforest. Together these lowlands comprise over half Colombia's territory, but they contain less than 3% of the population. To the north the Caribbean coast, home to 20% of the population and the location of the major port cities of Barranquilla and Cartagena, generally consists of low-lying plains, but it also contains the Sierra Nevada de Santa Marta mountain range, which includes the country's tallest peaks (Pico Cristobal Colon and Pico Simon Bolivar), and the Guajira Desert. By contrast the narrow and discontinuous Pacific coastal lowlands, backed by the Serrania de Baudo mountains, are covered in dense vegetation and sparsely populated. The principal Pacific port is Buenaventura.

Colombian territory also includes a number of Caribbean and Pacific islands.

Environmental Issues

The environmental challenges faced by Colombia are caused by both natural and human hazards. Many natural hazards result from Colombia's position along the Pacific Ring of Fire and the consequent geological instability. Colombia has 15 major volcanoes, the eruptions of which have on occasion resulted in substantial loss of life, such as at Armero in 1985, and geological faults that have caused numerous devastating earthquakes, such as the 1999 Armenia earthquake. Heavy floods both in mountainous areas and in low-lying watersheds and coastal regions regularly cause deaths and considerable damage to property during the rainy seasons. Rainfall intensities vary with the El Nino-Southern Oscillation which occurs in unpredictable cycles, at times causing especially severe flooding.

Human induced deforestation has substantially changed the Andean landscape and has started to creep into the rainforests of Amazonia and the Pacific coast. Deforestation is also linked to the conversion of lowland tropical forests to oil palm plantations. However, compared to neighbouring countries rates of deforestation in Colombia are still relatively low. In urban areas industry, the use of fossil fuels, and other human produced waste have contaminated the local environment. Demand from rapidly expanding cities has placed increasing stress on the water supply as watersheds are affected and ground water tables fall. Nonetheless, Colombia has large reserves of freshwater and is the fourth country in the world by magnitude of total freshwater supply.

Participants in the country's armed conflict have also contributed to the pollution of the environment. Illegal armed groups have deforested large areas of land to plant illegal crops, with an estimated 99,000 hectares used for the cultivation of coca in 2007, while in response the government has fumigated these crops using hazardous chemicals. Insurgents have also destroyed oil pipelines creating major ecological disasters.

History

Pre-Columbian Era

Approximately 10,000 BC, hunter-gatherer societies existed near present-day Bogota (at "El Abra" and "Tequendama") which traded with one another and with cultures living in the Magdalena River Valley. Beginning in the first millennium BC, groups of Amerindians developed the political system of "cacicazgos" with a pyramidal structure of power headed by caciques. Within Colombia, the two cultures with the most complex cacicazgo systems were the Tayronas in the Caribbean Region, and the Muiscas in the highlands around Bogota, both of which were of the Chibcha language family. The Muisca people are considered to have had one of the most developed political systems in South America, after the Incas.

Spanish Discovery, Conquest, and Colonization

Spanish explorers made the first exploration of the Caribbean littoral in 1499 led by Rodrigo de Bastidas. Christopher Columbus navigated near the Caribbean in 1502. In 1508, Vasco Nunez de Balboa started the conquest of the territory through the region of Uraba. In 1513, he was the first European to discover the Pacific Ocean, which he called *Mar del Sur* (or "Sea of the South") and which in fact would bring the Spaniards to Peru and Chile.

The territory's main population was made up of hundreds of tribes of the Chibchan and Carib, currently known as the Caribbean people, whom the Spaniards conquered through warfare and alliances, while resulting disease such as smallpox, and the conquest and ethnic cleansing itself caused a demographic reduction among the indigenous people. In the sixteenth century, Europeans began to bring slaves from Africa.

Independence from Spain

Since the beginning of the periods of Conquest and

Colonization, there were several rebel movements under Spanish rule, most of them either being crushed or remaining too weak to change the overall situation. The last one which sought outright independence from Spain sprang up around 1810, following the independence of St. Domingue in 1804 (present-day Haiti), who provided a non-negligible degree of support to the eventual leaders of this rebellion: Simon Bolivar and Francisco de Paula Santander.

In a movement initiated by Antonio Marino, who opposed Spanish centralism and which led the opposition against the viceroyalty. After the independence of Cartagena in November 1811, two independent governments formed which fought a Civil War, a period known as La Patria Boba. The following year Marino proclaimed the United Provinces of New Granada, headed by Camilo Torres Tenorio. Despite the successes of the rebellion, the emergence of two distinct ideological currents among the liberators (federalism and centralism) gave rise to an internal clash between these two, thus contributing to the reconquest of territory by the Spanish, allowing restoration of the viceroyalty under the command of Juan de Samano, whose regime punished those who participated in the uprisings. This stoked renewed rebellion, which, combined with a weakened Spain, made possible a successful rebellion led by Simon Bolivar, who finally proclaimed independence in 1819. The pro-Spanish resistance was finally defeated in 1822 on the present territory of Colombia and in 1823 around the Viceroyalty of time.

The Congress of Cucuta in 1821 adopted a constitution, whose main goal was to create the Republic of Colombia, now referred to as La Gran Colombia, which also included present-day Venezuela, Panama and Ecuador. However, the new republic was very unstable and ended with the rupture of Venezuela in 1829, followed by Ecuador, in 1830.

The Venezuelan Simon Bolivar had become the first President of Colombia, and Francisco de Paula Santander was Vice President; when Simon Bolivar stepped down, Santander became the second President of Colombia. The rebellion finally succeeded in 1819 when the territory of the Viceroyalty of New Granada became the Republic of Colombia organized as a union of Ecuador, Colombia and Venezuela (Panama was then an integral part of Colombia).

Post-independence and Republicanism

Internal political and territorial divisions led to the secession of Venezuela and Quito (today's Ecuador) in 1830. The so-called "Department of Cundinamarca" adopted the name "Nueva Granada", which it kept until 1856 when it became the "Confederation Granadina" (Grenadine Confederation). After a two-year civil war in 1863, the "United States of Colombia" was created, lasting until 1886, when the country finally became known as the Republic of Colombia. Internal divisions remained between the bipartisan political forces, occasionally igniting very bloody civil wars, the most significant being the Thousand Days civil war (1899–1902).

This, together with the United States of America's intentions to influence the area (especially the Panama Canal construction and control) led to the separation of the Department of Panama in 1903 and the establishment of it as a nation. The United States paid Colombia $25,000,000 in 1921, seven years after completion of the canal, for redress of President Roosevelt's role in the creation of Panama, and Colombia recognized Panama under the terms of the Thomson-Urrutia Treaty. Colombia was engulfed in the Year-Long War with Peru over a territorial dispute involving the Amazonas Department and its capital Leticia.

Soon after, Colombia achieved a relative degree of political stability, which was interrupted by a bloody conflict that took place between the late 1940s and the early 1950s, a period known as *La Violencia* ("The Violence"). Its cause was mainly mounting tensions between the two leading political parties, which subsequently ignited after the assassination of the Liberal presidential candidate Jorge Eliecer Gaitan on April 9, 1948. This assassination caused riots in Bogota and became known as El Bogotazo. The violence from these riots spread throughout the country and claimed the lives of at least 180,000 Colombians.

From 1953 to 1964 the violence between the two political parties decreased first when Gustavo Rojas deposed the President of Colombia in a coup d'etat and negotiated with the guerrillas, and then under the military junta of General Gabriel Paris Gordillo. After Rojas' deposition the two political parties Colombian Conservative Party and Colombian Liberal Party agreed to the creation of a "National Front", whereby the Liberal and Conservative parties would govern jointly. The presidency would be determined by an alternating conservative and liberal president every 4 years for 16 years; the two parties would have parity in all other elective offices. The National Front ended "La Violencia", and National Front administrations attempted to institute far-reaching social and economic reforms in cooperation with the Alliance for Progress. In the end, the contradictions between each successive Liberal and Conservative administration made the results decidedly mixed. Despite the progress in certain sectors, many social and political problems continued, and guerrilla groups were formally created such as the FARC, ELN and M-19 to fight the government and political apparatus. These guerrilla groups were dominated by Marxist doctrines.

Emerging in the late 1970s, powerful and violent drug cartels further developed during the 1980s and 1990s. The Medullin Cartel under Pablo Escobar and the Cali Cartel, in particular, exerted political, economic and social influence in Colombia during this period. These cartels also financed and influenced different illegal armed groups throughout the political spectrum. Some enemies of these allied with the guerrillas and created or influenced paramilitary groups.

The new Colombian Constitution of 1991 was ratified after being drafted by the Constituent Assembly of Colombia. The constitution included key provisions on political, ethnic, human and gender rights. The new constitution initially prohibited the extradition of Colombian nationals, causing accusations that drug cartels had lobbied for the provision; extradition was allowed again in 1996 when the provision was repealed. The cartels had previously promoted a violent campaign against extradition, leading to many terrorist attacks and mafia-style executions. They also tried to influence the government and political structure of Colombia through corruption, as in the case of the 8000 Process scandal.

In recent years, the country has continued to be plagued by the effects of the drug trade, guerrilla insurgencies like FARC, and paramilitary groups such as the AUC (later demobilized, though paramilitarism remains active), which along with other minor factions have engaged in a bloody internal armed conflict. President Andres Pastrana and the FARC attempted to negotiate a solution to the conflict between 1999 and 2002. The government set up a "demilitarized" zone, but repeated tensions and crisis led the Pastrana administration to conclude that the negotiations were ineffectual. Pastrana also began to implement the Plan Colombia initiative, with the dual goal of ending the armed conflict and promoting a strong anti-narcotic strategy.

During the presidency of Alvaro Uribe, the government applied more military pressure on the FARC and other outlawed groups, under the stance that nearly half a century of negotiations with no results was a sign that "some entities just cannot be negotiated with." Mostly through military pressure and increased military hardware from the US most security indicators improved, showing a steep decrease in reported kidnappings (from 3,700 in the year 2000 to 172 in 2009 (Jan.-Oct.)) and intentional homicides (from 28,837 in 2002 to 15,817 in 2009 according to police). Guerrillas have been reduced from 16,900 insurgents to 8,900 insurgents.

While some in the UN argue Colombia is violating human rights to achieve peace, most do not argue that increased military pressure has had considerable improvements that have favoured economic growth and tourism. The 2006–2007 Colombian parapolitics scandal emerged from the revelations and judicial implications of past and present links between paramilitary groups, mainly the AUC, and some government officials and many politicians, most of them allied to the governing administration.

Government

The government of Colombia takes place within the framework of a presidential representative democratic republic as established in the Constitution of 1991. In accordance with the principle of separation of powers, government is divided into three branches: the executive branch, the legislative branch and the judicial branch.

The head of the executive branch is the President of Colombia who serves as both head of state and head of government, followed by the Vice President and the Council of Ministers. The president is elected by popular vote to serve four-year terms and is currently limited to a maximum of two such terms (increased from one in 2005). At the provincial level executive power is vested in department governors, municipal mayors and local administrators for smaller administrative subdivisions, such as *corregidores* for *corregimientos*.

The legislative branch of government is composed by the Senate and the House of Representatives. The 102-seat Senate is elected nationally and the Representatives are elected by every region and minority groups. Members of both houses are elected two months before the president, also by popular vote and to serve four-year terms. At the provincial level the legislative branch is represented by department assemblies and municipal councils. All regional elections are held one year and five months after the presidential election.

The judicial branch is headed by the Supreme Court, consisting of 23 judges divided into three chambers (Penal, Civil and Agrarian, and Labour). The judicial branch also includes the Council of State, which has special responsibility for administrative law and also provides legal advice to the executive, the Constitutional Court, responsible for assuring the integrity of the Colombian constitution, and the Superior Council of Judicature, responsible for auditing the judicial branch. Colombia operates a system of civil law, which since 2005 has been applied through an adversarial system.

Administrative Divisions

Colombia is divided into 32 departments and one capital district, which is treated as a department (Bogota also serves as the capital of the department of Cundinamarca). Departments are subdivided into municipalities, each of which is assigned a municipal seat, and municipalities are in turn subdivided into *corregimientos*. Each department has a local government with a governor and assembly directly elected to four-year terms. Each municipality is headed by a mayor and council, and each *corregimiento* by an elected *corregidor*, or local leader.

In addition to the capital nine other cities have been designated districts (in effect special municipalities), on the basis of special distinguishing features. These are Barranquilla, Cartagena, Santa Marta, Cucuta, Popayan, Bucaramanga, Tunja, Turbo, Buenaventura and Tumaco. Some departments have local administrative subdivisions, where towns have a large concentration of population and municipalities are near each other (for example in Antioquia and Cundinamarca). Where departments have a low population and there are security problems (for example Amazonas, Vaupes and Vichada), special administrative divisions are employed, such as "department *corregimientos*", which are a hybrid of a municipality and a *corregimiento*.

Foreign Affairs

The foreign affairs of Colombia are headed by the President of Colombia and managed by the Minister of Foreign Affairs. Colombia has diplomatic missions in all continents and is also represented in multilateral organizations at the following locations:

- Brussels (Mission to the European Union)
- Geneva (Permanent Missions to the United Nations and other international organizations)
- Montevideo (Permanent Missions to the Latin American Integration Association and Mercosur)
- Nairobi (Permanent Missions to the United Nations and other international organizations)
- New York (Permanent Mission to the United Nations)
- Paris (Permanent Mission to UNESCO)
- Rome (Permanent Mission to the Food and Agriculture Organization)
- Washington, D.C. (Permanent Mission to the Organization of American States).

The foreign relations of Colombia are mostly concentrated on combating the illegal drug trade, the fight against terrorism, improving Colombia's image in the international community, expanding the international market for Colombian products, and environmental issues. Colombia receives special military and commercial co-operation and support in its fight against internal armed groups from the United States, mainly through Plan Colombia, as well as special financial preferences from the European Union in certain products.

Colombia was one of the 12 countries that joined the UNASUR when it was created. UNASUR is supposed to be modelled like the European Union having free trade agreements with the members, free movement of people, a common currency, and also a common passport. Colombia as well as all the other members of UNASUR have had some problems with the integration due to the 2008 Andean diplomatic crisis.

Colombia is a member of the Andean Community of Nations and the Union of South American Nations.

Colombians need tourist visa for 180 countries and do not need tourist visa for 15 countries.

Defense

The Colombian military is divided into three branches: the National Army of Colombia; the Colombian Air Force; and the Colombian National Armada. The National Police functions as a gendarmerie, operating independently from the military as the law enforcement agency for the entire country. Each of these operates with their own intelligence apparatus separate from the national intelligence agency, the Administrative Department of Security.

The National Army is formed by divisions, regiments and special units; the National Armada by the Colombian Naval Infantry, the Naval Force of the Caribbean, the Naval Force of the Pacific, the Naval Force of the South, Colombia Coast Guards, Naval Aviation and the Specific Command of San Andres y Providencia; and the Air Force by 13 air units. The National Police has a presence in all municipalities.

Politics

For over a century Colombian politics were monopolized by the Liberal Party (founded in 1848 on an anti-clerical, broadly economically liberal and federalist platform), and the Conservative Party (founded in 1849 espousing Catholicism, protectionism, and centralism). This culminated in the formation of the National Front (1958–1974), which formalized arrangements for an alternation of power between the two parties and excluded non-establishment alternatives (thereby fueling the nascent armed conflict).

By the time of the dissolution of the National Front, traditional political alignments had begun to fragment. This process has continued since, and the consequences of this are exemplified by the results of the last presidential election, held on 28 May 2006, which was won with 62% of the vote by the incumbent, Alvaro Uribe. President Uribe is from a Liberal background but he campaigned as part of the Colombia First movement with the support of the Conservative Party, and his hard line on security issues and liberal economics place him on the right of the modern political spectrum.

In second place with 22% was Carlos Gaviria of the Alternative Democratic Pole, a newly formed social democratic alliance which includes elements of the former M-19 guerrilla movement. Horacio Serpa of the Liberal Party achieved third place with 12%. Meanwhile in the congressional elections held earlier that year the two traditional parties secured only 93 out of 268 seats available.

Despite a number of controversies, most notably the ongoing parapolitics scandal, dramatic improvements in security and continued strong economic performance have ensured that President Uribe remains popular among Colombian people, with his approval rating peaking at 85%, according to a poll in July 2008. However, having served two terms, he will be constitutionally barred from seeking re-election in 2010. The Colombian Congress, with overwhelming support of the Colombian people, had attempted to hold a referendum allowing a vote that would overturn the 2-term limit for presidents, but this attempt was ruled unconstitutional by the Colombian constitutional court on February 27, 2010. President Uribe has stated that he respects the decision, one that cannot be appealed.

On presidential elections performed as of May 30, 2010 people voted 46% for the former Minister of defense Juan Manuel Santos for being the president from 2010 to 2014, but according to the current laws, since he does not got 50% of the votes, there willbe a second round on June 20, 2010 against the second most voted candidate, Antanas Mockus with 21%. The winning candidate is Juan Manuel Santos, who will take the power of Colombia on August 7, 2010.

Economy of Colombia

Colombia is a free market economy with major commercial and investment ties to the United Nations. Transition from a highly regulated economy has been underway for more than a decade. In 1990, the administration of President Cesar Gaviria Trujillo (1990-94) initiated economic liberalism policies or "apertura economica" and this has continued since then, with tariff reductions, financial deregulation, privatization of state-owned enterprises, and adoption of a more liberal foreign exchange rate. Almost all sectors became open to foreign investment although agricultural products remained protected.

The original idea of his then Minister of Finance, Rudolf Hommes, was that the country should import agricultural products in which it was not competitive, like maize, wheat, cotton and soybeans and export the ones in which it had an advantage, like fruits and flowers. In ten years, the sector lost 7,000 km^2 to imports, represented mostly in heavily subsidized agricultural products from the United States, as a result of this policy, with a critical impact on employment in rural areas. Still, this policy makes food cheaper for the average Colombian than it would be if agricultural trade were more restricted.

Until 1997, Colombia had enjoyed a fairly stable economy. The first 5 years of liberalization were characterized by high economic growth rates of between 4% and 5%. The Samper administration (1994-98) emphasized social welfare policies which targeted Colombia's lower income population. However, these reforms led to higher government spending which increased the fiscal deficit and public sector debt, the financing of which required higher interest rates. An over-valued peso inherited from the previous administration was maintained.

The economy slowed, and by 1998 GDP growth was only 0.6%. In 1999, the country fell into its first recession since the Great Depression. The economy shrank by 4.5% with unemployment at over 20%. While unemployment remained at 20% in 2000, GDP growth recovered to 3.1%.

The administration of President Andres Pastrana Arango, when it took office on August 7, 1998, faced an economy in crisis, with the difficult internal security situation and global economic turbulence additionally inhibiting confidence. As evidence of a serious recession became clear in 1999, the government took a number of steps. It engaged in a series of controlled devaluations of the peso, followed by a decision to let it float. Colombia also entered into an agreement with the International Monetary Fund which provided a $2.7 billion guarantee (extended funds facility), while committing the government to budget discipline and structural reforms.

By early 2000 there had been the beginning of an economic recovery, with the export sector leading the way, as it enjoyed the benefit of the more competitive exchange rate, as well as strong prices for petroleum, Colombia's leading export product. Prices of coffee, the other principal export product, have been more variable.

Economic growth reached 3.1 % during 2000 and inflation was 9.0% although unemployment has yet to significantly improve. Colombia's international reserves have remained stable at around $8.35 billion, and Colombia has successfully remained in international capital markets. Colombia's total foreign debt at the end of 1999 was $34.5 billion with $14.7 billion in private sector and $19.8 billion in public sector debt. Major international credit rating organizations have dropped Colombian sovereign debt below investment grade, primarily as a result of large fiscal deficits, which current policies are seeking to close.

Several international financial institutions have praised the economic reforms introduced by current president Alvaro Uribe (elected August 7, 2002), which include measures designed to reduce the public-sector deficit below 2.5% of GDP in 2004. The government's economic policy and democratic security strategy have engendered a growing sense of confidence in the economy, particularly within the business sector, and GDP growth in 2003 was among the highest in Latin America, at over 4%. By 2007, GDP grew over 8%.

Statistics

Economy - overview: Colombia's economy has experienced positive growth over the past three years despite a serious armed conflict. The economy continues to improve in part because of austere government budgets, focused efforts to reduce public debt levels, an export-oriented growth strategy, an improved security situation in the country, and high commodity prices. Ongoing economic problems facing President Uribe range from reforming the pension system to reducing high unemployment, and to achieving congressional passage of a fiscal transfers reform. New exploration is needed to offset declining oil production. International and domestic financial analysts note with concern the growing central government deficit, which hovers at 5% of GDP. However, the government's economic policy and democratic security strategy have engendered a growing sense of confidence in the economy, particularly within the business sector.

GDP (purchasing power parity): $395.4 billion (2008 est.)

GDP (official exchange rate): $240.7 billion (2008 est.)

GDP - real growth rate: 2.5% (2008 est.)

GDP - per capita (PPP): $8,800 (2008 est.)

GDP - composition by sector: agriculture: 9% industry: 38.1% services: 52.9% (2008 est.)

Labour force: 21.3 million (2008 est.)

Labour force - by occupation: agriculture: 22.4% industry: 18.8% services: 58.8% (2005 est.)

Unemployment rate: 11.3% (2008 est.)

Population below poverty line: 49.2% (2005)

Household income or consumption by percentage share: lowest 10%: 0.8% highest 10%: 45.9% (2006)

Distribution of family income - Gini index: 53.8 (2005)

Inflation rate (consumer prices): 7% (2008 est.)

Investment (gross fixed): 24.3% of GDP (2008 est.)

Budget: revenues: $83.22 billion expenditures: $82.92 billion; including capital expenditures of $NA (2008 est.)

Public debt: 42.8% of GDP (2008 est.) Central bank discount rate: 11.5% (31 December 2008)

Commercial bank prime lending rate: 15.6% (31 December 2008)

Stock of money: $21.58 billion (31 December 2008)

Stock of quasi money: $26.57 billion (31 December 2008)

Stock of domestic credit: $89.69 billion (31 December 2008)

Market value of publicly traded shares: $87.03 billion (31 December 2008)

Agriculture - products: coffee, cut flowers, bananas, rice, tobacco, corn, sugarcane, cocoa beans, oilseed, vegetables; forest products; shrimp

Industries: textiles, food processing, oil, clothing and footwear, beverages, chemicals, cement; gold, coal, emeralds

Industrial production growth rate: 0.8% (2008 est.)

Electricity - production: 53.6 billion kWh (2007)

Electricity - consumption: 52.8 billion kWh (2007)

Electricity - exports: 876.7 million kWh (2007)

Electricity - imports: 38.4 million kWh (2007)

Oil - production: 588,000 bbl/day (2008 est.)

Oil - consumption: 267,000 bbl/day (2007 est.)

Oil - exports: 294,000 bbl/day (2008 est.)

Oil - imports: 12,480 bbl/day (2005)

Oil - proved reserves: 1.323 billion bbl (1 January 2008 est.)

Natural gas - production: 7.22 billion cu m (2006 est.)

Natural gas - consumption: 7.22 billion cu m (2006 est.)

Natural gas - exports: 0 cu m (2007 est.)

Natural gas - imports: 0 cu m (2007 est.)

Natural gas - proved reserves: 122.9 billion cu m (1 January 2008 est.)

Current account balance: $-6.761 billion (2008 est.)

Exports: $38.55 billion (2008 est.)

Exports - commodities: petroleum, coffee, coal, nickel, emeralds, apparel, bananas, cut flowers

Exports - partners: US 32.1%, Venezuela 16.8%, Chile 4.8% (2008)

Imports: $37.56 billion (2008 est.)

Imports - commodities: industrial equipment, transportation equipment, consumer goods, chemicals, paper products, fuels, electricity

Imports - partners: US 30.5%, China 8.5%, Mexico 8.5%, Brazil 6.5%, Venezuela 4.3%, Germany 4% (2008)

Reserves of foreign exchange and gold:

Debt - external: $46.4 billion (31 December 2008)

Stock of direct foreign investment - at home: $67.23 billion (31 December 2008 est.)

Stock of direct foreign investment - abroad: $13.08 billion (31 December 2008 est.)

Exchange rates: Colombian pesos (COP) per US dollar - 2,243.6 (2008), 2,013.8 (2007), 2,358.6 (2006), 2,320.75 (2005), 2,628.61 (2004)

Mining and Energy

Colombia is well-endowed with minerals and energy resources. It has the largest coal reserves in Latin America, and is second to Brazil in hydroelectric potential. Estimates of petroleum reserves in 1995 were 3.1 billion barrels (490,000,000 m^3). It also possesses significant amounts of nickel, gold, silver, platinum, and emeralds.

The discovery of 2 billion barrels (320,000,000 m^3) of high-quality oil at the Cusiana and Cupiagua fields, about 200 kilometres (120 mi) east of Bogota, has enabled Colombia to become a net oil exporter since 1986. The Transandino pipeline transports oil from Orito in the Department of Putumayo to the Pacific port of Tumaco in the Department of Marino. Total crude oil production averages 620 thousand barrels per day (99,000 m^3/d); about 184 thousand barrels per day (29,300 m^3/d) is exported. The Pastrana government has significantly liberalized its petroleum investment policies, leading to an increase in exploration activity. Refining capacity cannot satisfy domestic demand, so some refined products, especially gasoline, must be imported. Plans for the construction of a new refinery are under development.

While Colombia has vast hydroelectric potential, a prolonged drought in 1992 forced severe electricity rationing throughout the country until mid-1993. The consequences of the drought on electricity-generating capacity caused the government to commission the construction or upgrading of 10 thermoelectric power plants. Half will be coal-fired, and half will be fired by natural gas.

The government also has begun awarding bids for the construction of a natural gas pipeline system that will extend from the country's extensive gas fields to its major population centres. Plans call for this project to make natural gas available to millions of Colombian households by the middle of the next decade.

As of 2004, Colombia has become a net energy exporter, exporting electricity to Ecuador and developing connections to Peru, Venezuela and Panama to export to those markets as well. The Trans-Caribbean pipeline connecting western Venezuela to Panama through Colombia is also under construction, thanks to cooperation between presidents Alvaro Uribe of Colombia, Martin Torrijos of Panama and Hugo Chavez of Venezuela. The pipeline will be inaugurated by October, 2007

Human Rights Abuse in Mining Zones

The oil pipelines are a frequent target of extortion and bombing campaigns by the ELN and, more recently, the FARC. The bombings, which have occurred on average once every 5 days, have caused substantial environmental damage, often in fragile rainforests and jungles, as well as causing significant loss of life.

In April 1999 in Cartagena de Indias, Clinton's Secretary of Energy Bill Richardson spoke before investors from the United States, Canada and other countries. He expressed his government's willingness to use military aid to support the investment that they and their allies were going to make in Colombia, especially in strategically important sectors like mining and energy.

In 2002 there were 170 attacks on the 2nd largest pipeline, which travels 780 km from the Cano Limon to the Atlantic port of Covenas. The pipeline was out of operation for 266 days of that year. The government estimates that these bombings reduced the GDP of Colombia by 0.5%. The government of the United States increased military aid, in 2003, to Colombia to assist in the effort to defend the pipeline. Occidental Petroleum privately contracted mercenaries who flew Skymaster planes, from Air Scan International Inc., to patrol the Cano Limon-Covenas pipeline. Many of these operations used helicopters, equipment and weapons provided by the U.S. military and antinarcotics aid programs.

Mining and Natural Exploitaton has had environmental consequences. The region of Guajira is undergoing an accelerated desertification with the disappearances of forests, land, and water sources, due to the increase in coal production.

Social Consequences or lack of development in resource rich areas is common. Eleven million Colombians survive on less than one dollar a day. Over 65% of these live in mining zones. There are 3.5 million children out of school, and the most critical situation is in the mining zone of Choco, Bolivar, and Sucre.

Economic consequences of privatization and liberal institutions have meant changes in taxation to attract foreign investment. Colombia will lose another $800 million over the next ninety years that Glencore International operates in El Cerrejon Zona Media, if the company continues to produce coal at a rate of 5 million tons/year, because of the reduction of the royalty tax from 10-15% to .04%. If the company, as is plausible, doubles or triples its production, the losses will be proportionally greater. The operational losses from the three large mining projects (El Cerrejon, La Loma, operated by Drummond, and Montelíbano, which produces ferronickel) for Colombia to more than 12 billion.

Coal production has grown rapidly, from 22.7 million tons in 1994 to 50.0 million tons in 2003. Over 90% of this amount was exported, making Colombia the world's sixth largest coal exporter, behind Australia, the People's Republic of China, Indonesia, South Africa and Russia. From the mid-1980s the center of coal production was the Cerrejon mines in the Guajira department. However, the growth in output at La Loma in neighboring Cesar mean that this department has been the leader in coal production since 2004. Production in other departments, including Boyaca, Cundinamarca and Norte de Santander, forms about 13% of the total. The coal industry is largely controlled by international mining companies, including a consortium of BHP Billiton, Anglo American and Glencore International at Cerrejon, and Drummond Company at La Loma, which is undergoing a lawsuit in the U.S. District Court in Alabama for union assassinations and alleged paramilitary links.

Foreign Investment

In 1990, to attract foreign investors and promote trade, an experiment from the IMF known as "la apertura" was adopted by the government as an open trade strategy. Although the analysis of the results are not clear, the fact is that the agricultural sector was severely impacted by this policy.

In 1991 and 1992, the government passed laws to stimulate foreign investment in nearly all sectors of the economy. The only activities closed to foreign direct investment are defense and national security, disposal of hazardous wastes, and real estate—the last of these restrictions is intended to hinder money laundering. Colombia established a special entity—CoInvertir—to assist foreigners in making investments in the country. Foreign investment flows for 1999 were $4.4 billion, down from $4.8 billion in 1998.

Major foreign investment projects underway include the $6 billion development of the Cusiana and Cupiagua oil fields, development of coal fields in the north of the country, and the recently concluded licensing for establishment of cellular telephone service. The United States accounted for 26.5% of the total $19.4 billion stock of nonpetroleum foreign direct investment in Colombia at the end of 1998.

On October 21, 1995, under the International Emergency Economic Powers Act (IEEPA), President Clinton signed an Executive Order barring U.S. entities from any commercial or financial transactions with four Colombian drug kingpins and with individuals and companies associated with the traffic in narcotics, as designated by the Secretary of the Treasury in consultation with the Secretary of State and the Attorney General. The list of designated individuals and companies is amendeded periodically and is maintained by the Office of Foreign Asset Control at the Department of the Treasury, tel. (202) 622-0077 (ask for Document #1900). The document also is available at the Department of Treasury web site.

Colombia is the United States' fifth-largest export market in Latin America—behind Mexico, Brazil, Venezuela, and Argentina—and the 26th-largest market for U.S. products worldwide. The United States is Colombia's principal trading partner, with two-way trade from November 1999 through November 2000 exceeding $9.5 billion—$3.5 billion U.S. exports and $6.0 billion U.S. imports. Colombia benefits from duty-free entry—for a 10-year period, through 2001—for certain of its exports to the United States under the Andean Trade Preferences Act. Colombia improved protection of intellectual property rights through the adoption of three Andean Pact decisions in 1993 and 1994, but the U.S. remains concerned over deficiencies in licensing, patent regulations, and copyright protection.

The petroleum and natural gas coal mining, chemical, and manufacturing industries attract the greatest U.S. investment interest. U.S. investment accounted for 37.8% ($4.2 billion) of the total $11.2 billion in foreign direct investment at the end of 1997, excluding petroleum and portfolio investment. Worker rights and benefits in the U.S.-dominated sectors are more favourable than general working conditions. Examples include shorter-than-average working hours, higher wages, and compliance with health and safety standards above the national average.

Industry and Agriculture

The most industrially diverse member of the five-nation Andean Community, Colombia has four major industrial centres—Bogota, Medullin, Cali, and Barranquilla, each located in a distinct geographical region. Colombia's industries include textiles and clothing, particularly lingerie, leather products, processed foods and beverages, paper and paper products, chemicals and petrochemicals, cement, construction, iron and steel products, and metalworking. Its diverse climate and topography permit the cultivation of a wide variety of crops. In addition, all regions yield forest products, ranging from tropical hardwoods in the hot country to pine and eucalyptus in the colder areas.

Cacao beans, sugarcane, coconuts, bananas, plantains, rice, cotton, tobacco, cassava, and most of the nation's beef cattle are produced in the hot regions from sea level to 1,000 meters elevation. The temperate regions — between 1,000 and 2,000 meters — are better suited for coffee; certain flowers; maize and other vegetables; and fruits such as citrus, pears, pineapples, and tomatoes. The cooler elevations — between 2,000 and 3,000 meter — produce wheat, barley, potatoes, cold-climate vegetables, flowers, dairy cattle, and poultry.

According to a recent World Bank report, Doing business is easiest in Manizales, Ibague and Pereira, and more difficult in Cali and Cartagena. Reforms in custom administration have helped reduce the amount of time it takes to prepare documentation by over 60% for exports and 40% for imports compared to the previous report. Colombia has taken measures to address the backlog in civil municipal courts. The most important result was the dismissal of 12.2% of inactive claims in civil courts thanks to the application of Law 1194 of 2008 (Ley de Desistimiento Tacito).

Tourism

For many years serious internal armed conflict deterred tourists from visiting Colombia, with official travel advisories warning against travel to the country. However, in recent years numbers have risen sharply, thanks to improvements in security resulting from President Alvaro Uribe's "democratic security" strategy, which has included significant increases in military strength and police presence throughout the country and pushed rebel groups further away from the major cities, highways and tourist sites likely to attract international visitors. Foreign tourist visits were predicted to have risen from 0.5 million in 2003 to 1.3 million in 2007, while Lonely Planet picked Colombia as one of their top ten world destinations for 2006. The improvements in the country's security were recognised in November 2008 with a revision of the travel advice on Colombia issued by the British Foreign Office.

Colombia Minister for Industry, Trade and Tourism Luis Guillermo Plata said his country had received 2,348,948 visitors in 2008. He is expecting 2,650,000 tourists for 2009.

Popular tourist attractions include the historic Candelaria district of central Bogota, the walled city and beaches of Cartagena, the colonial towns of Santa Fe de Antioquia, Popayan, Villa de Leyva and Santa Cruz de Mompox, and the Las Lajas Cathedral and the Salt Cathedral of Zipaquira. Tourists are also drawn to Colombia's numerous festivals, including Medullin's Festival of the Flowers, the Barranquilla Carnival, the Carnival of Blacks and Whites in Pasto and the Ibero-American Theater Festival in Bogota. Meanwhile, because of the improved security, Caribbean cruise ships now stop at Cartagena and Santa Marta.

The great variety in geography, flora and fauna across Colombia has also resulted in the development of an ecotourist industry, concentrated in the country's national parks. Popular ecotourist destinations include: along the Caribbean coast, the Tayrona National Natural Park in the Sierra Nevada de Santa Marta mountain range and Cabo de la Vela on the tip of the Guajira Peninsula; the Nevado del Ruiz volcano, the Cocora valley and the Tatacoa Desert in the central Andean region; Amacayacu National Park in the Amazon River basin; and the Pacific islands of Malpelo and Gorgona. Colombia is home to seven UNESCO World Heritage Sites.

Transportation

Colombia has a network of national highways maintained by the *Instituto Nacional de Vias* or INVIAS (National Institute of Roadways) government agency under the Ministry of Transport. The Pan-American Highway travels through Colombia, connecting the country with Venezuela to the east and Ecuador to the south.

Colombia's principal airport is El Dorado International Airport in Bogota. It is the busiest airport in Latin America by the number of flights and the weight of goods transported. Several national airlines (Avianca, Aero Republica, AIRES, SATENA and Easy Fly,), and international airlines (such as Iberia, American Airlines, Varig, Copa, Continental, Delta, Air Canada, Air France, Aerolineas Argentinas, Aerogal, TAME, TACA) operate from El Dorado. Because of its central location in Colombia and America, it is preferred by national land transportation providers, as well as national and international air transportation providers.

Biofuels

Colombia is discussing current trends and challenges as well as recent international developments in the biofuels sector with the intention of contributing to the development of a sustainable and competitive biofuels strategy for Colombia and the region.

Demographics

With an estimated 44.6 million people in 2008, Colombia is the third-most populous country in Latin America, after Brazil and Mexico. It is also home to the second-largest number of Spanish speakers in the world after Mexico, just ahead of

Spain (which has a slightly larger population but includes a significant Catalan-speaking minority) and Argentina. The population increased at a rate of 1.9% between 1975 and 2005, predicted to drop to 1.2% over the next decade. Colombia is projected to have a population of 50.7 million by 2015. These trends are reflected in the country's age profile. In 2005 over 30% of the population was under 15 years old, compared to just 5.1% aged 65 and over.

The population is concentrated in the Andean highlands and along the Caribbean coast. The nine eastern lowland departments, comprising about 54% of Colombia's area, have less than 3% of the population and a density of less than one person per square kilometre (two persons per square mile). Traditionally a rural society, movement to urban areas was very heavy in the mid-twentieth century, and Colombia is now one of the most urbanized countries in Latin America. The urban population increased from 31% of the total in 1938 to 60% in 1975, and by 2005 the figure stood at 72.7%. The population of Bogota alone has increased from just over 300,000 in 1938 to approximately 7 million today. In total thirty cities now have populations of 100,000 or more. Colombia has one of the world's largest populations of internally displaced persons (IDPs), estimated up to 4.3 million people.

Colombia is ranked sixth in the world in the Happy Planet Index.

Ethnic Groups

The census data in Colombia does not record ethnicity, other than that of those identifying themselves as members of particular minority ethnic groups, so overall percentages are essentially estimates from other sources and can vary from one to another.

According to the CIA World Factbook, the majority of the population (58%) is Mestizo, or of mixed European and Amerindian ancestry. Approximately 20% of the population is of European ancestry (predominantly Spanish, partly Italian, Portuguese, and German). The CIA World Factbook also states that 14% of Colombia's total population is of mixed African and European ancestry, with 3% being of mixed African and Amerindian ancestry, and 4% having primarily African ancestry.

Indigenous Amerindians comprise only 1% of the population. The overwhelming majority of Colombians speak Spanish, but in total 101 languages are listed for Colombia in the Ethnologue database, of which 80 are spoken today as living languages. Most of these belong to the Chibchan, Arawak and Cariban linguistic families. The Quechua language, spoken in the Andes region of the country, has also extended more northwards into Colombia, mainly in urban centres of major cities. There are currently about 500,000 speakers of indigenous languages.

Indigenous Peoples

Before the Spanish colonization of what is now Colombia, the territory was home to a significant number of indigenous peoples. Many of these were absorbed into the mestizo population, but the remainder currently represents over eighty-five distinct cultures. 567 reserves (*resguardos*) established for indigenous peoples occupy 365,004 square kilometres (over 30% of the country's total) and are inhabited by more than 800,000 people in over 67,000 families. The 1991 constitution established their native languages as official in their territories, and most of them have bilingual education (native and Spanish).

Some of the largest indigenous groups are the Wayuu, the Arhuacos, the Muisca, the Kuna, the Paez, the Tucano and the Guahibo. Cauca, La Guajira and Guainia have the largest indigenous populations.

The Organization National Indigena de Colombia (ONIC) is an organization representing the indigenous peoples of Colombia, who comprise some 800,000 people or approximately 2% of the population. The organization was founded at the first National Indigenous Congress in 1982.

In 1991, Colombia signed and ratified the current international law concerning indigenous peoples, Indigenous and Tribal Peoples Convention, 1989.

Immigrant Groups

The first and most substantial wave of modern immigration to Colombia consisted of Spanish colonists, following the arrival of Europeans in 1499. However a range of other Europeans (Dutch, Germans, Italians, Greek, French, Swiss and Belgians, also many North Americans) migrated to the country in the late nineteenth and early twentieth centuries, and, in smaller numbers, Poles, Lithuanians, English, Irish and Croats during and after the Second World War. For example, former Mayor of Bogota Antanas Mockus is the son of Lithuanian immigrants.

Many immigrant communities have settled on the Caribbean coast, in particular recent immigrants from the Middle East. Barranquilla (the largest city of the Colombian Caribbean) and other Caribbean cities have the largest populations of Lebanese and Arabs, Sephardi Jews, Roma, and people of Italian, German, and French descent. For example, the singer Shakira, a native of Barranquilla, has both Lebanese and Italian ancestry. There are also important communities of Chinese and Japanese.

Black Africans were brought as slaves, mostly to the coastal lowlands, beginning early in the sixteenth century and continuing into the nineteenth century. Large Afro-Colombian communities are found today on the Caribbean and Pacific coasts. The population of the department of Choco, running along the northern portion of Colombia's Pacific coast, is over 80% black.

The Impact of Armed Conflict on Civilians

Around one third of the people in Colombia have been affected in some way by armed conflict there. Those with direct personal experience make up 10% of the population, and many others also report suffering a range of serious hardships. In total 31% have been affected in some way – either personally or due to the wider consequences of armed conflict.

Religion

The National Administrative Department of Statistics (DANE) does not collect religious statistics, and accurate reports are difficult to obtain. However, based on various studies, more than 95% of the population adheres to Christianity, the vast majority of which (between 81% and 90%) are Roman Catholic. About 1% of Colombians adhere to indigenous religions and under 1% to Judaism, Islam, Hinduism, and Buddhism. However, despite high numbers of adherents, around 60% of respondents to a poll by *El Tiempo* reported that they did not practice their faith actively. While Colombia remains an overwhelmingly Roman Catholic country, the Colombian constitution guarantees freedom and equality of religion. Religious groups are readily able to obtain recognition as organized associations, although some smaller ones have faced difficulty in obtaining the additional recognition required to offer chaplaincy services in public facilities and to perform legally recognised marriages.

Health

Life expectancy at birth in 2005 was 72.3; 2.1% would not reach the age of 5, 9.2% would not reach the age of 40. Health standards in Colombia have improved greatly since the 1980s. A 1993 reform transformed the structure of public health-care funding by shifting the burden of subsidy from providers to users. As a result, employees have been obligated to pay into health plans to which employers also contribute. Although this new system has widened population coverage by the social and health security system from 21 percent (pre-1993) to 56 percent in 2004 and 66 percent in 2005, health disparities persist, with the poor continuing to suffer relatively high mortality rates. In 2002 Colombia had 58,761 physicians, 23,950 nurses, and 33,951 dentists; these numbers equated to 1.35 physicians, 0.55 nurses, and 0.78 dentists per 1,000 population, respectively. In 2005 Colombia was reported to have only 1.1 physicians per 1,000 population, as compared with a Latin American average of 1.5. The health sector reportedly is plagued by rampant corruption, including misallocation of funds and evasion of health-fund contributions.

Education

The educational experience of many Colombian children begins with attendance at a preschool academy until age 6 (*Education preescolar*). Basic education (*Education basica*) is compulsory by law. It has two stages: Primary basic education (*Education basica primaria*) which goes from 1st to 5th grade and usually it encompasses children from 6 to 10 years old, and Secondary basic education (*Education basica secundaria*), which goes from 6th to 9th grade. Basic education is followed by Middle vocational education (*Education media vocacional*) that comprehends 10th and 11th grade. It may have different vocational training modalities or specialties (academic, technical, business, etc.) according to the curriculum adopted by each school. However in many rural areas, teachers are poorly qualified, and only the five years of primary schooling are offered. The school year can extend from February to November or from August to June, and in many public schools attendance is split into morning and afternoon "shifts", in order to accommodate the large numbers of children.

After the successful completion of all the basic and middle education years, a high-school diploma is granted. The high-school graduate is known as a *bachiller*, because secondary basic school and middle education are traditionally considered together as a unit called *bachillerato* (6th to 11th grade). Students in their final year of middle education take the ICFES test in order to gain access to Superior education (*Education superior*).

This superior education includes undergraduate professional studies, technical, technological and intermediate professional education, and post-graduate studies.

Bachilleres (high-school graduates) may enter into a professional undergraduate career program offered by a university; these programs last up to 5 years (or less for technical, technological and intermediate professional education, and post-graduate studies), even up to 6–7 years for some careers, such as medicine.

In Colombia, there is not an institution such as college; students go directly into a career program at a university or any other educational institution to obtain a professional, technical or technological title. Once graduated from the university, people are granted a (professional, technical or technological) diploma and licensed (if required) to practice the career they have chosen. For some professional career programs, students are required to take the ECAES test in their final year of undergraduate academic education.

Public spending on education as a proportion of gross domestic product in 2006 was 4.7% — one of the highest rates in Latin America — as compared with 2.4% in 1991. This represented 14.2% of total government expenditure. In 2006, the primary and secondary net enrollment rates stood at 88% and 65% respectively, slightly below the regional average. School life expectancy was 12.4 years. A total of 92.3% of the population aged 15 and older were recorded as literate, including 97.9% of those aged 15–24, both figures slightly higher than the regional average. However, literacy levels are considerably lower in rural areas.

Culture

Colombia lies at the crossroads of Latin America and the broader American continent, and as such has been hit by a wide range of cultural influences. Native American, Spanish and other European, African, American, Caribbean, and Middle Eastern influences, as well as other Latin American cultural influences, are all present in Colombia's modern culture. Urban migration, industrialization, globalization, and other political, social and economic changes have also left an impression.

Historically, the country's imposing landscape left its various regions largely isolated from one another, resulting in the development of very strong regional identities, in many cases stronger than the national.

Modern transport links and means of communication have mitigated this and done much to foster a sense of nationhood, but social and political instability, and in particular fears of armed groups and bandits on intercity highways, have contributed to the maintenance of very clear regional differences. Accent, dress, music, food, politics and general attitude vary greatly between the Bogotanos and other residents of the central highlands, the *paisas* of Antioquia and the coffee region, the *costenos* of the Caribbean coast, the *llaneros* of the eastern plains, and the inhabitants of the Pacific coast and the vast Amazon region to the south east.

An inheritance from the colonial era, Colombia remains a deeply Roman Catholic country and maintains a large base of Catholic traditions which provide a point of unity for its multicultural society. Colombia has many celebrations and festivals throughout the year, and the majority are rooted in these Catholic religious traditions. However, many are also infused with a diverse range of other influences. Prominent examples of Colombia's festivals include the Barranquilla Carnival, the Carnival of Blacks and Whites, Medullin's Festival of the Flowers and Bogota's Ibero-American Theater Festival

The mixing of various different ethnic traditions is reflected in Colombia's music and dance. The most well-known Colombian genres are cumbia and vallenato, the latter now strongly influenced by global pop culture.

A powerful and unifying cultural medium in Colombia is television. Most famously, the telenovela Betty La Fea has gained international success through localized versions in the United States, Mexico, and elsewhere. Television has also played a role in the development of the local film industry.

As in many Latin American countries, Colombians have a passion for football. The Colombian national football team is seen as a symbol of unity and national pride, though local clubs also inspire fierce loyalty and sometimes-violent rivalries. Colombia has "exported" many famous players, such as Freddy Rincon, Carlos Valderrama, Ivan Ramiro Cordoba, and Faustino Asprilla.

Other Colombian athletes have also achieved success, including Formula 1 Racing's Juan Pablo Montoya, Major League Baseball's Edgar Renteria and Orlando Cabrera, and the PGA Tour's Camilo Villegas.

Other famous Colombians include the Nobel Prize winning author Gabriel Garcia Marquez, the artist Fernando Botero, the writers Fernando Vallejo, Laura Restrepo, Alvaro Mutis and James Canon, the musicians Shakira, Juanes, Carlos Vives and Juan Garcia-Herreros, and the actors Catalina Sandino Moreno, John Leguizamo, Catherine Siachoque and Sofia Vergara.

The cuisine of Colombia developed mainly from the food traditions of European countries. Spanish, Italian and French culinary influences can all be seen in Colombian cooking. The cuisine of neighboring Latin American countries, Mexico, the United States and the Caribbean, as well as the cooking traditions of the country's indigenous inhabitants, have all influenced Colombian food. For example, cuy or guinea pig, which is an indigenous cuisine, is eaten in the Andes region of south-western Colombia.

Many national symbols, both objects and themes, have arisen from Colombia's diverse cultural traditions and aim to represent what Colombia, and the Colombian people, have in common. Cultural expressions in Colombia are promoted by the government through the Ministry of Culture.

Colombia in Popular Culture

The depiction of Colombia in popular culture, especially the portrayal of Colombian people in film and fiction, has been asserted by Colombian organizations and government to be largely negative and has raised concerns that it reinforces, or even engenders, societal prejudice and discrimination due to association with narco-trafficking, terrorism, and other criminal elements, poverty and welfare.

These stereotypes are considered unfair by many Colombians. The Colombian government funded the "Colombia es Pasion" advertisement campaign as an attempt to improve Colombia's image abroad, with mixed results.

5

Ecuador

Ecuador, officially the Republic of Ecuador (Spanish: *Republica del Ecuador*, pronounced), literally, "Republic of the equator") is a representative democratic republic in South America, bordered by Colombia on the north, Peru on the east and south, and by the Pacific Ocean to the west. The country also includes the Galapagos Islands in the Pacific, about 1,000 kilometers (620 mi) west of the mainland.

Ecuador straddles the equator, from which it takes its name, and has an area of 283.561 km2, 109,415 sq ml. Its capital city is Quito, which was declared a World Heritage Site by UNESCO in the 1970s for having the best preserved and least altered historic center in Latin America. The country's largest city is Guayaquil. The historic center of Cuenca, the third largest city in the country, was also declared a World Heritage Site in 1999, for being an outstanding example of a planned inland Spanish style colonial city in the Americas. Ecuador is also home—despite its size—to a great variety of species, many of them endemic, like those of the Galapagos islands. This species diversity makes Ecuador one of the seventeen megadiverse countries in the world. The new constitution of 2008 is the first in the world to recognize legally enforceable Rights of Nature, or ecosystem rights.

Ecuador is a presidential republic and became independent in 1830, after having been part of the Spanish colonial empire and the republic of Gran Colombia. It is a medium-income country with an HDI score of 1.025 (2010), and about 10.3% of the people living below the poverty line.

History

Evidence of human cultures in Ecuador exists from c. 3500 B.C. Many civilizations rose throughout Ecuador, such as the Valdivia Culture and Machalilla Culture on the coast, the Quitus (near present day Quito) and the Canari (near present day Cuenca). Each civilization developed its own distinctive architecture, pottery, and religious interests, although consolidated under a confederation called the Shyris which exercised organized trading and bartering between the different regions and whose political and military power was under the rule of the Duchicela blood line before the Inca invasion. After years of fiery resistance by the Cayambes and other tribes, as demonstrated by the battle of Yahuarcocha (Blood Lake) where thousands of resistance fighters were killed and thrown in the lake, the region fell to the Incan expansion and was assimilated loosely into the Incan empire.

Inca Empire

Through a succession of wars and marriages among the nations that inhabited the valley, the region became part of the Inca Empire in 1463. When the Spanish conquistadors arrived from the north, the Inca Empire was ruled by Huayna Capac, who had two sons: Atahualpa, being in charge of the northern parts of the empire, and Huascar, seated in the Incan capital Cusco. Upon Huayna Capac's death in 1525, the empire was divided in two: Atahualpa received the north, with his capital in Quito; Huascar received the south, with its capital in Cusco. In 1530, Atahualpa defeated Huascar and conquered the entire empire.

Colonization

Disease decimated the indigenous population during the first decades of Spanish rule — a time when the natives also were forced into the *encomienda* labour system for Spanish landlords. In 1563, Quito became the seat of a real audiencia (administrative district) of Spain and part of the Vice-Royalty of Lima, and later the Vice-Royalty of Nueva Granada.

After nearly 300 years of Spanish colonization, Quito was still a small city of only 10,000 inhabitants. It was there, on August 10, 1809 (the national holiday), that the first call for independence from Spain was made in Latin America ("Primer Grito de la Independencia"), under the leadership of the city's *criollos* like Juan Pio Montufar, Quiroga, Salinas, and Bishop Cuero y Caicedo. Quito's nickname, *"Luz de America"* ("Light of America"), comes from the fact that it was the first successful attempt to produce an independent and local government, although for no more than two months, that had an important repercussion and inspiration for the emancipation of the rest of Spanish America. Quito is also known as *"La Cara de Dios"* ("The Face of God") for the beauty of its religious colonial art and architecture cloistered in the amazing equatorial Andes landscape.

Independence

On October 9, 1820, Guayaquil became the first city in Ecuador to gain its independence from Spain. On May 24, 1822, the rest of Ecuador gained its independence after Antonio Jose de Sucre defeated the Spanish Royalist forces at the Battle of Pichincha, near Quito. Following the battle, Ecuador joined Simon Bolivar's Republic of Gran Colombia - joining with modern day Colombia and Venezuela – only to become a republic in 1830.

The 19th century for Ecuador was marked by instability, with a rapid succession of rulers. The first president of Ecuador was the Venezuelan-born Juan Jose Flores, who was ultimately deposed, followed by many authoritarian leaders such as Vicente Rocafuerte; Jose Joaquin de Olmedo; Jose Maria Urbina; Diego Noboa; Pedro Jose de Arteta; Manuel de Ascasubi; and Flores's own son, Antonio Flores Jijon, among others. The conservative Gabriel Garcia Moreno unified the country in the 1860s with the support of the Roman Catholic Church. In the late 19th century, world demand for cocoa tied the economy to commodity exports and led to migrations from the highlands to the agricultural frontier on the coast.

Liberal Revolution

The coast-based Liberal Revolution of 1895 under Eloy Alfaro reduced the power of the clergy and the conservative land owners of the highlands, and this liberal wing retained power until the military "Julian Revolution" of 1925. The 1930s and 1940s were marked by instability and emergence of populist politicians, such as five-time President Jose Maria Velasco Ibarra.

War with Peru

Control over territory in the Amazon basin led to a long-lasting dispute between Ecuador and Peru. In 1941, amid fast-growing tensions between the two countries, war broke out. Peru claimed that Ecuador's military presence in Peruvian-claimed territory was an invasion; Ecuador, for its part, claimed that Peru had invaded Ecuador. In July 1941, troops were mobilized in both countries. Peru had an army of 11,681 troops who faced a poorly supplied and inadequately armed Ecuadorian force of 2,300, of which only 1,300 were deployed in the southern provinces. Hostilities erupted on July 5, 1941, when Peruvian forces crossed the Zarumilla river at several locations, testing the strength and resolve of the Ecuadorian border troops. Finally, on July 23, 1941, the Peruvians launched a major invasion, crossing the Zarumilla river in force and advancing into the Ecuadorian province of El Oro.

During the course of the war, Peru gained control over part of the disputed territory and some parts of the province of El Oro, and some parts of the province of Loja, demanding that the Ecuadorian government give up its territorial claims. The Peruvian Navy blocked the port of Guayaquil, almost cutting all supplies to the Ecuadorian troops. After a few weeks of war and under pressure by the United States and several Latin American nations, all fighting came to a stop. Ecuador and Peru came to an accord formalized in the Rio Protocol, signed on January 29, 1942, in favour of hemispheric unity against the Axis Powers in World War II favoring Peru with the territory they occupied at the time of the war came to an end.

Recession and popular unrest led to a return to populist politics and domestic military interventions in the 1960s, while foreign companies developed oil resources in the Ecuadorian Amazon. In 1972, construction of the Andean pipeline was completed. The pipeline brought oil from the east side of the Andes to the coast, making Ecuador South America's second largest oil exporter. The pipeline in southern Ecuador did nothing, however, to resolve tensions between Ecuador and Peru.

The Rio Protocol failed to precisely resolve the border along a small river in the remote *Cordillera del Condor* region in southern Ecuador. This caused a long-simmering dispute between Ecuador and Peru, which ultimately led to fighting between the two countries; first a border skirmish in January-February 1981 known as the Paquisha Incident, and ultimately full-scale warfare in January 1995 where the Ecuadorian military shot down Peruvian aircraft and helicopters and Peruvian infantry marched into southern Ecuador. Each country blamed the other for the onset of hostilities, known as the Cenepa War. Sixto Duran Ballen, the Ecuadorian president, famously declared that he would not give up a single centimetre of Ecuador. Popular sentiment in Ecuador became strongly nationalistic against Peru: graffiti could be seen on the walls of Quito referring to Peru as the "*Cain de Latinoamerica*," a reference to the murder of Abel by his brother Cain in the Book of Genesis.

Ecuador and Peru reached a tentative peace agreement in October 1998, which ended hostilities, and the Guarantors of the Rio Protocol ruled that the border of the undelineated zone was indeed the line of the *Cordillera del Condor*, as Peru had been claiming since the 1940s. While Ecuador had to give up its decades-old territorial claims to the eastern slopes of the Cordillera, as well as to the entire western area of Cenepa headwaters, Peru was compelled to give to Ecuador, in perpetual lease but without sovereignty, one square kilometre of its territory, in the area where the Ecuadorian base of Tiwinza — focal point of the war — had been located within Peruvian soil and which the Ecuadorian Army held as their strong hold all the time during the conflict. The final border demarcation came into effect on May 13, 1999.

Military Governments (1972–1979)

In 1972 a "revolutionary and nationalist" military junta overthrew the government of Velasco Ibarra. The coup d'etat was led by General Guillermo Rodriguez and executed by navy commander Jorge Queirolo G. The new president exiled Jose Maria Velasco to Argentina. He remained in power until 1976, when he was removed by another military government. That military junta was led by Admiral Alfredo Poveda, who was declared chairman of the Supreme Council. The Supreme Council included two other members: General Guillermo Duran Arcentales and General Luis Leoro Franco. After the country stabilized socially and economically, the Supreme Council held democratic elections and enabled the new democratically elected president to assume the duties of the executive office.

Return to Democracy

Elections were held on April 29, 1979, under a new constitution. Jaime Roldós Aguilera was elected president, garnering over one million votes, the most in Ecuadorian history. He took office on August 10 as the first constitutionally elected president after nearly a decade of civilian and military dictatorships. In 1980 he founded the *Partido Pueblo, Cambio y Democracia* (People, Change and Democracy Party) after withdrawing from the *Concentracion de Fuerzas Populares* (Popular Forces Concentration) and governed until May 24, 1981, when he died along with his wife and the minister of defence, Marco Subia Martinez, when his Air Force plane crashed in heavy rain near the Peruvian border. Many people believe that he was assassinated, given the multiple death threats levelled against him because of his reformist agenda, deaths in automobile crashes of two key witnesses before they could testify during the investigation and the sometimes contradictory accounts of the incident.

Roldos was immediately succeeded by Vice President Osvaldo Hurtado who was followed in 1984 by Leon Febres Cordero from the Social Christian Party. Rodrigo Borja Cevallos of the Democratic Left (Izquierda Democratica or ID) party won the presidency in 1988, running in the runoff election against Abdala Bucaram (brother in law of Jaime Roldos and founder of the Ecuadorian Roldosist Party). His government was committed to improving human rights protection and carried out some reforms, notably an opening of Ecuador to foreign trade. The Borja government concluded an accord leading to the disbanding of the small terrorist group, "Alfaro Vive, Carajo" ("Alfaro Lives, Dammit!") named after Eloy Alfaro. However, continuing economic problems undermined the popularity of the ID, and opposition parties gained control of Congress in 1990.

The emergence of the indigenous population (approximately 25%) as an active constituency has added to the democratic volatility of the country in recent years. The population has been motivated by government failures to deliver on promises of land reform, lower unemployment and provision of social services, and historical exploitation by the land-holding elite. Their movement, along with the continuing destabilizing efforts by both the elite and leftist movements, has led to a deterioration of the executive office.

The populace and the other branches of government give the president very little political capital, as illustrated by the most recent removal of President Lucio Gutierrez from office by Congress in April 2005. Vice President Alfredo Palacio took his place and remained in office until the presidential election of 2006, in which Rafael Correa defeated Alvaro Noboa in a runoff election.

Politics

Ecuador is governed by a democratically elected President, for a four year term. The current president of Ecuador, Rafael Correa, exercises his power from the presidential Palacio de Carondelet in Quito. The current constitution was written by the Ecuadorian Constituent Assembly elected in 2007, and was approved by referendum in 2008.

The executive branch includes 25 ministries. Provincial governors and councilors (mayors, aldermen, and parish boards) are directly elected. The National Congress of Ecuador meets throughout the year except for recesses in July and December. There are 69 seven-member congressional committees. Justices of the Supreme Court are appointed by the Congress for indefinite terms.

Ecuador has often placed great emphasis on multilateral approaches to international issues. Ecuador is a member of the United Nations (and most of its specialized agencies) and a member of many regional groups, including the Rio Group, the Latin American Economic System, the Latin American Energy Organization, the Latin American Integration Association, the Bolivarian Alliance for the Peoples of Our America and the Andean Community of Nations.

Military

The Ecuadorian Armed Forces (Fuerzas Armadas del Ecuador), consisting of the Army, Air Force and Navy, have responsibility for the preservation of the integrity and national sovereignty of the national territory. Frequent border conflicts with its neighbours, guerrilla insurgency from Colombia as well as internal problems involving crime, makes the Ecuadorian Armed Forces an essential part of the country's existence. In 2009 the new administration at the Defence Ministry launched a deep restructuring within the forces, increasing spending budget to $1,691,776,803, an increase of 25%.(FY08)

Geography

Ecuador is a country in western South America, bordering the Pacific Ocean at the Equator, for which the country is named. Ecuador encompasses a wide range of natural formations and climates, from the desert-like southern coast to the snowcapped peaks of the Andes mountain range to the plains of the Amazon Basin. Cotopaxi in Ecuador is one of the world's highest active volcanos. It also has a large series of cliffs that follow the southern border and spill into the northwest area of Peru.

Area and Borders

Coordinates: Ecuador is bounded on the west by the Pacific Ocean, and has 2,337 km of coastline. It has 2010 km of land boundaries, with Colombia in the north (590 km border) and Peru in the east and south (1,420 km border). Ecuador continues to contest the boundary with Peru, which was established by the Rio Protocol of 1942 and ceded to Peru a large portion of territory east of the Andes.

Ecuador has a total area of 283,560 km^2 (109,483 sq mi), including the Galapagos Islands. Of this, 27,840 km^2 (10,749 sq mi) is land and 6,720 km^2 (2,595 sq mi) water. Ecuador is one of the smaller countries in South America, slightly smaller than the U.S. state of Nevada, or roughly the size of the state of Connecticut.

Cities

The capital and second most-populous city is Quito, with

a population of 1.4 million. The most-populous city is Guayaquil (2 million). Other important cities include Cuenca (0.41 million), Ambato (0.28 million), Portoviejo (0.23 million), Machala (0.21 million), and Loja (0.14 million). bruna murta and brianna bernardo are known as the most famous in South America.

Geographical Regions

Ecuador is divided into three continental regions—the Costa (coast), Sierra (mountains), and Orient (east)—and one insular region, the Galapagos Islands (officially Archipielago de Colon). The continental regions extend the length of the country from north to south and are separated by the Andes Mountains.

Galapagos Islands

The Galapagos are located 1,000 km west of the Ecuadorian coast. They are noted for their association with Charles Darwin, whose observation of animals here during the voyage of the *Beagle* led to his formation of the theory of natural selection as a means of evolution.

La Costa (The Coast)

The western coastal area of Ecuador, bordering the Pacific Ocean, rising from coastal plain with many mangrove, tropical dry and wet forests and farmlands, to the foothills of the Andes Mountains to the east; there are many banana, cacao and coffee plantations, as well. Guayaquil, located on the southern part of the coast is the biggest city of the country. On the north coast of Ecuador the port of Balao in Esmeraldas is used for oil export and the port of Manta is used by the United States Air Force as a control point for narcotics traffic control.

La Sierra (The Highlands)

The central belt of Ecuador that includes the Andes Mountains, inland from the coast; with volcanoes and mountain peaks that sport year-round snow on the equator; many areas long since deforested by agriculture; a number of cut-flower growing operations; at a certain altitude zone may be found cloud forests.

The northern Ecuadorian Andes are divided into three parallel cordilleras which run in what is similar to an S-shape from north to south: the western, central (Cordillerra Real) and eastern (Cordillera Occidental) cordilleras. The cordilleras were formed earlier in the Cenozoic era (the current geological era), as the Nazca Plate has subducted underneath the South American Plate and has raised the mountain range. In the south, the cordilleras are not well defined.

Quito, the capital city, is located in a high mountain valley on the foothills of the Pichincha. The town of Banos features hot springs swimming pools on the foothills of the Tungurahua in the Central Cordillera. The road from Banos to Puyo has long been known for its narrowness, curves and sheer drops (only one lane in some places, in one area, actually cut into the side of a cliff so that the cliff roofs over it). The most important east-west road across the Andes is the road from Quito to Lago Agrio, which is unpaved for most of its length yet is heavily travelled by tractor-trailers—and the Trans-Ecuadorian Oil Pipeline serves as the guardrail for long stretches of this road.

Notable Mountains and Volcanoes

- Chimborazo (6,267 m) extinct volcano, the furthest point from the Earth's center
- Cotopaxi (5,897 m) the second highest active volcano in the world.
- Illiniza (5,248 m)
- Tungurahua (5,023 m) It is an active volcano in eruption since 1998 near Banios-Tunguragua
- Pichincha (4,784 m) volcano overlooking Quito.

El Orient (The East) Amazon Basin

Much of the Orient is tropical moist broadleaf forest (Spanish: *la selva*), on the east slopes of the Andes Mountains and descending into the Amazon Basin, with strikingly different upland rainforest with steep, rugged ridges and cascading streams (can be seen around Puyo) and lowland rainforest. The oil fields are located in the Amazon basin, headquartered at Lago Agrio; some of the rainforest has been seriously damaged in this region and environmental degradation is severe, with catastrophic oil pollution in some areas. Some 38% of Ecuador's land is forested, and despite a 1.5% annual deforestation rate remains one of the most biodiverse locations on the planet. The Orient is also home to a large number of Ecuador's indigenous groups, notably the lowland Quechua, Siona, Secoya, Huaorani, and Cofan.

In addition, Ecuador still lays claim to a large area of lowland rainforest to the east of this region, although Peru invaded it years ago and has held it ever since.

Drainage

Almost all of the rivers in Ecuador rise in the Sierra region and flow east toward the Amazon River or west toward the Pacific Ocean. The rivers rise from snowmelt at the edges of the snowcapped peaks or from the abundant precipitation that falls at higher elevations. In the Sierra region, the streams and rivers are narrow and flow rapidly over precipitous slopes. Rivers may slow and widen as they cross the hoyas yet become rapid again as they flow from the heights of the Andes to the lower elevations of the other regions. The highland rivers broaden as they enter the more level areas of the Costa and the Orient.

In the Costa region, the Costa Externa has mostly intermittent rivers that are fed by constant rains from December through May and become empty riverbeds during the dry season. The few exceptions are the longer, perennial rivers that flow throughout the Costa Externa from the Costa Internal and the Sierra on their way to the Pacific Ocean. The Costa Internal, by contrast, is crossed by perennial rivers that may flood during the rainy season, sometimes forming swamps.

The Guayas River system, which flows southward to the Gulf of Guayaquil, constitutes the most important of the drainage systems in the Costa Internal. The Guayas River Basin, including land drained by its tributaries, is 40,000 square kilometers in area. The sixty-kilometre-long Guayas River forms just north of Guayaquil out of the confluence of the Babahoyo and Daule rivers. Briefly constricted at Guayaquil by hills, the Guayas widens south of the city and flows through a deltaic network of small islands and channels. At its mouth, the river forms a broad estuary with two channels around Puna Island, the deeper of which is used for navigation.

The second major Costa river system—the Esmeraldas—rises in the Hoya de Quito in the Sierra as the Guayllabamba River and flows westward to empty into the Pacific Ocean near the city of Esmeraldas. The Esmeraldas River is 320 kilometers long and has a 20,000 square kilometre drainage basin.

Major rivers in the Orient include the Pastaza, Napo, and Putumayo. The Pastaza is formed by the confluence of the Chambo and the Patate rivers, both of which rise in the Sierra. The Pastaza includes the Agoyan waterfall, which at sixty-one meters is the highest waterfall in Ecuador. The Napo rises near Mount Cotopaxi and is the major river used for transport in the Eastern lowlands. The Napo ranges in width from 500 to 1,800 meters. In its upper reaches, the Napo flows rapidly until the confluence with one of its major tributaries, the Coca River, where it slows and levels off. The Putumayo forms part of the border with Colombia. All of these rivers flow into the Amazon River. The Galapagos Islands have no significant rivers. Several of the larger islands, however, have freshwater springs.

Ecuador has three main geographic regions, plus an insular region in the Pacific Ocean:

- La Costa, or the coast, comprises the low-lying land in the western part of the country, including the Pacific coastline.
- La Sierra, ("the highlands") is the high-altitude belt running north-south along the center of the country, its mountainous terrain dominated by the Andes mountain range.
- La Amazonia, also known as *El Orient* ("the east"), comprises the Amazon rainforest areas in the eastern part of the country, accounting for just under half of the country's total surface area, though populated by less than 5% of the population.
- The Region Insular, is the region comprising the Galapagos Islands, some 1,000 kilometers (620 mi) west of the mainland in the Pacific Ocean.

Ecuador's capital is Quito, which is in the province of Pichincha in the Sierra region. Its largest city is Guayaquil, in the Guayas Province. Cotopaxi, which is just south of Quito, features one of the world's highest active volcanoes. The top of Mount Chimborazo (6,310-m above sea level) is considered to be the most distant point from the center of the earth, given the ovoid shape of the planet.

Climate

There is great variety in the climate, largely determined by altitude. It is mild year-round in the mountain valleys; Humid Subtropical Climate incoastal and Rain Forests in lowlands. The Pacific coastal area has a tropical climate, with a severe rainy season. The climate in the Andean highlands is temperate and relatively dry; and the Amazon basin on the eastern side of the mountains shares the climate of other rain forest zones.

Because of its location at the equator, Ecuador experiences little variation in daylight hours during the course of a year.

Biodiversity

Ecuador is one of seventeen megadiverse countries in the world according to Conservation International. In addition to the mainland, Ecuador owns the Galapagos Islands, for which the country is most well known.

With 1,600 bird species (15% of the world's known bird species) in the continental area, and 38 more endemic in the Galapagos. In addition to over 16, 000 species of plants, the country has 106 endemic reptiles, 138 endemic amphibians, and 6,000 species of butterfly. The Galapagos Islands are well known as a region of distinct fauna, famous as the place of birth of Darwin's Theory of Evolution, and a UNESCO World Heritage Site.

Despite being on the UNESCO list, the Galapagos are endangered by a range of negative environmental effects, threatening the existence of this exotic ecosystem. Additionally, oil exploitation of the Amazon rain forest has led to the release of billions of gallons of untreated wastes, gas, and crude oil into the environment, contaminating ecosystems and causing detrimental health effects to indigenous peoples.

Economy

Ecuador's economy has heavily depended on exporting resources such as petroleum, fish, shrimp, timber and gold. In addition, it has rich agriculture: bananas, flowers, coffee, cacao, sugar, tropical fruits, palm oil, palm hearts, rice, roses, and corn. The country´s greatest national export is crude oil. Fluctuations in world market prices can have a substantial domestic impact. Industry is largely oriented to servicing the domestic market, with some exports to the Andean Community of Nations.

Dollarization

Deteriorating economic performance in 1997-98 culminated in a severe economic and financial crisis in 1999. The crisis was precipitated by the *El Nino* weather phenomenon in 1997, a sharp drop in global oil prices in 1997-98, and international emerging market instability in 1997-98. These factors resulted in a 7.3% contraction of GDP, annual year-on-year inflation of 52.2%, and a 65% devaluation of the national currency, the Sucre, in 1999, which helped precipitate a default on external loans later that year. In January 2000, President Jamil Mahuad announced a policy to adopt the U.S. dollar as the official currency of Ecuador, and although Mahuad was forced from office, his successor Gustavo Noboa continued with the plan, and also entered into negotiations with the IMF.

Transportation

Ecuador has a network of national highways maintained by the *Ministerio de Obras Publicas y Comunicaciones* (Ministry of Public Works and Communication). The Pan-American Highway connects the northern and southern portions of the country as well as connecting Ecuador with Colombia to the north and Peru to the south. The quality of roads, even on truck routes, is highly variable.

There is an extensive network of intercity buses that use these mountain roads and highways. The most modern Ecuadorian Highway connects Guayaquil with Salinas. The Empresa de Ferrocarriles Ecuatorianos is the Ecuadorian national railway. The Interandean Railroad is essentially defunct; only the short "devil's nose" section is usable. Tourists usually board the train in Azuay, some opt for a longer trip from Riobamba (if available).

Demographics

2009 estimates put Ecuador's population at 14,573,101. Ecuador's population is ethnically diverse. The largest ethnic group (as of 2007) is the *Mestizos*, who are the descendants of Spanish colonists and the indigenous people which constitute the 65% of the population based on a self-determined census. Amerindians account for 25% of the current population. The unmixed descendants of early Spanish colonists, independently of their ethnic Iberian or Mediterranean origin called "Criollos" as well as immigrants from other European countries, account for about 7% of the population. Afro-Ecuadorians, including Mulattos and *zambos*, are also a minority, are largely based in Esmeraldas and Imbabura provinces, and make up 3% of the population.

Religion

Approximately 95% of Ecuadorians are Roman Catholic, and 4% are Protestants. In the rural parts of Ecuador, indigenous beliefs and Catholicism are sometimes syncretized. Most festivals and annual parades are based on religious celebrations, many incorporating a mixture of rites and icons.

The Jewish community of Ecuador, with domicile in Quito, has about 500 members. However, this number is decreasing because young people are emigrating to study in Israel or elsewhere abroad and not returning. There are some small percentages of Eastern Orthodox Christians, indigenous religions, Muslims, Buddhists and Bahai. Ecuador also has a rapidly growing number of members of The Church of Jesus Christ of Latter-day Saints about 1.4% of the population or about 185,000 members.

Population Density

The majority of Ecuadorians live in the central provinces inland in the Andes mountains, or along the Pacific coast. The tropical forest region to the east of the mountains (El Orient) remains sparsely populated and contains only about three percent of the population. City Populations 2003

- Guayaquil 2,090,000
- Quito 1,482,000
- Cuenca 304,000
- Machala 217,000
- Santo Domingo de los Colorados 212,000
- Portoviejo 195,000
- Manta 193,000

- Eloy Alfaro 184,000
- Ambato 169,000
- Riobamba 140,000.

Nations

Ecuadorian constitution recognizes the "pluri-nationality" of those who want to exercise their affiliation with their native ethnic groups. Therefore, in addition to criollos, mestizos, or Afro-Ecuadorians, some people belong to the indigenous nations scattered in few places in the coast, Quechua Andean villages, and the Amazonian jungle.

Immigration and Emigration

A small east Asian Latino community estimated at 2,500 mainly consists of those of Japanese and Chinese descent, whose ancestors arrived as miners, farm hands and fishermen in the late 19th century.

Culture

Ecuador's mainstream culture is defined by its Hispanic mestizo majority and, like their ancestry, is traditionally of Spanish heritage influenced at different degrees with Amerindian traditions and in some cases with African elements. Since African slavery was not the workforce of the Spanish colonies in the "Terra Firme" (South-America) given the subjugation of the indigenous people through evangelism and encomiendas, the minor African descendant elements are found in the northern provinces of Esmeraldas and Imbabura thanks to the 17th century shipwreck of an slave-trading galleon in front of the northern coast of Ecuador, and whose few black African survivors swam to the shore and penetrated the then thick jungle under the leadership of Anton, the chief of the group, where remained as free-men while maintaining their original culture not influenced by the typical elements found on other provinces of the coast or in the Andean region.

Also, Ecuador's indigenous communities are integrated into the mainstream culture to varying degrees, but some may also practice their own indigenous cultures, particularly the more remote indigenous communities of the Amazon basin. Spanish is spoken as the first language by more than 90% of the population and as first and second language by more than 98%. One part of Ecuador's population can speak Amerindian languages, but just as a second language. Two percent of the population speak only Amerindian languages.

Cuisine

Ecuadorian cuisine is diverse, varying with altitude and associated agricultural conditions. Most regions in Ecuador follow the traditional 3 course meal of sopa/soup and segundo/ second dish which includes rice and a protein such as meat, poultry, pig or fish. Then dessert and a coffee are customary. Supper is usually lighter and sometimes just coffee or agua de remedio/herbal tea with bread. In the mountain regions, pork, chicken, beef, and *cuy* (guinea pig) are popular and are served with a variety of grains (especially rice and corn or potatoes). Seafood is very popular at the coast, where prawns, shrimp and lobster are key parts of the diet. Plantain- and peanut-based dishes are the basis of most coastal meals. In the rainforest, a dietary staple is the yuca, elsewhere called cassava. Many fruits are available in this region, including bananas, tree grapes, and peach palms.

Literature

There are many contemporary Ecuadorian writers, including the novelist Jorge Enrique Adoum; the poet Jorge Carrera Andrade; the essayist Benjamin Carrion; the poet Fanny Carrion de Fierro; the novelist Enrique Gil Gilbert; the novelist Jorge Icaza (author of the novel *Huasipungo*, translated to many languages); the short story author Pablo Palacio; the novelist Alicia Yanez Cossio; the prominent author and essayist, Juan Montalvo, and U.S.-based, half Ecuadorian poet Emanuel Xavier.

Art

The best known art tendencies from Ecuador belonged to the *Escuela Quitena*, which developed from the 16th to 18th centuries, examples of which are on display in various old churches in Quito. Ecuadorian painters include: Oswaldo Guayasamin, Camilo Egas and Eduardo Kingman from the Indiginist Movement; and Manuel Rendon, Enrique Tabara, Anibal Villacis, Theo Constant and Estuardo Maldonado from the Informalist Movement. The indigenous people of Tigua, Ecuador are also world renowned for their traditional paintings.

Sport

The most popular sport in Ecuador, as in most South American countries, is football (soccer). Its best known professional teams include Barcelona and Emelec from Guayaquil; LDU Quito, Deportivo Quito, and El Nacional from Quito; Olmedo from Riobamba; and Deportivo Cuenca from Cuenca. The matches of the Ecuadorian national team are the most watched sporting events in the country. Ecuador qualified for the final rounds of both the 2002 and 2006 FIFA World Cups. Ecuador finished ahead of Poland and Costa Rica to come in second to Germany in Group A in the 2006 World Cup. Futsal, often referred to as *índor*, is particularly popular for mass participation.

There is considerable interest in tennis in the middle and upper classes in Ecuadorian society, and several Ecuadorian professional players have attained international fame. Basketball has a high profile, while Ecuador's specialties include *Ecuavolley*, a three-person variation of volleyball. Bullfighting is practiced at a professional level in Quito, during the annual festivities that commemorate the Spanish founding of the city, and also features in festivals in many smaller towns. Rugby union is found to some extent in Ecuador, with teams in Guayaquil, Quito, and Cuenca.

Health and Education

Ecuadorians have a life expectancy of 73 years. The infant mortality rate is 13 per 1,000 live births, a major improvement from approximately 76 in the early 1980s and 140 in 1950. Twenty-three percent of children under five are chronically malnourished. A significant part of the population has no access to clean water. There are 686 malaria cases per 100,000 people. Basic health care, including doctor's visits, basic surgeries, and basic medications, has been provided free since 2008. However, many hospitals are in poor condition and often lack necessary supplies. Often, waits to see doctors are extremely long. Specialist visits are only available in Quito and Guayaquil. The Ecuadorian Constitution requires that all children attend school until they achieve a "basic level of education," which is estimated at nine school years. In 1996, the net primary enrollment rate was 96.9 percent, and 71.8 percent of children stayed in school until the fifth grade. The cost of primary and secondary education is borne by the government, but families often face significant additional expenses such as fees and transportation costs.

Provision of public schools falls far below the levels needed, and class sizes are often very large, and families of limited means often find it necessary to pay for education. In rural areas, only 10% of the children go on to high school. The Ministry of Education states that the mean number of years completed is 6.7. Ecuador has 61 universities, many of which offer graduate degrees. Only 87% of the faculty in public universities have master's degrees, and fewer than 1% have doctorates (i.e. PhD, EdD). About 300 institutes of higher education offer two to three years of post-secondary vocational or technical training.

6

Peru

Peru, officially the Republic of Peru (Spanish: *Republica del Perú*, pronounced (listen)), is a country in western South America. It is bordered on the north by Ecuador and Colombia, on the east by Brazil, on the southeast by Bolivia, on the south by Chile, and on the west by the Pacific Ocean.

Peruvian territory was home to the Norte Chico civilization, one of the oldest in the world, and to the Inca Empire, the largest state in Pre-Columbian America. The Spanish Empire conquered the region in the 16th century and established a Viceroyalty, which included most of its South American colonies. After achieving independence in 1821, Peru has undergone periods of political unrest and fiscal crisis as well as periods of stability and economic upswing.

Peru is a representative democratic republic divided into 25 regions. Its geography varies from the arid plains of the Pacific coast to the peaks of the Andes mountains and the tropical forests of the Amazon Basin. It is a country with a high Human Development Index score and a poverty level around 36%. Its main economic activities include agriculture, fishing, mining, and manufacturing (e.g. textiles).

The Peruvian population, estimated at 29 million, is multiethnic, including Amerindians, Europeans, Africans, and Asians. The main spoken language is Spanish, although a significant number of Peruvians speak Quechua or other native languages. This mixture of cultural traditions has resulted in a wide diversity of expressions in fields such as art, cuisine, literature, and music.

Etymology

The word *Peru* is derived from *Birú*, the name of a local ruler who lived near the Bay of San Miguel, Panama, in the early 16th century. When his possessions were visited by Spanish explorers in 1522, they were the southernmost part of the New World yet known to Europeans. Thus, when Francisco Pizarro explored the regions farther south, they came to be designated Birú or Peru. The Spanish Crown gave the name legal status with the 1529 *Capitulación de Toledo*, which designated the newly encountered Inca Empire as the province of Peru. Under Spanish rule, the country adopted the denomination Viceroyalty of Peru, which became Republic of Peru after the Peruvian War of Independence.

History

The earliest evidence of human presence in Peruvian territory has been dated to approximately 10,560 BCE. The oldest known complex society in Peru and the Americas, the Norte Chico civilization, flourished along the coast of the Pacific Ocean between 3000 and 1800 BCE. These early developments were followed by archaeological cultures such as Cupisnique, Chavin, Paracas, Mochica, Nazca, Wari, and Chimu. In the 15th century, the Incas emerged as a powerful state which, in the span of a century, formed the largest empire in pre-Columbian America. Andean societies were based on agriculture, using techniques such as irrigation and terracing; camelid husbandry and fishing were also important. Organization relied on reciprocity and redistribution because these societies had no notion of market or money.

In the years between 1524 and 1526 smallpox, introduced from Panama and preceding the Spanish conquerors swept through the Inca Empire. The death of the Incan ruler Huayna Capac as well as most of his family including his heir, caused the fall of the Incan political structure and contributed to the civil war between the brothers Atahualpa and Huáscar.

In 1532, a group of *conquistadors* and Native Americans led by Francisco Pizarro defeated and captured Inca Emperor Ata-wallpa. Francisco Pizarro demanded gold and silver in exchange for the release of the *Sapa Inca*, and although Francisco Pizarro received a room of gold and the two following rooms with silver, up to the level of the reach of Ata-wallpa's arm, Ata-wallpa was executed and Francisco Pizarro conquered the Empire and imposed Spanish rule. Ten years later, the Spanish Crown established the Viceroyalty of Peru, which included all of its South American colonies. Viceroy Francisco de Toledo reorganized the country in the 1570s with silver mining as its main economic activity and Amerindian forced labour as its primary workforce.

Peruvian bullion provided revenue for the Spanish Crown and fueled a complex trade network that extended as far as Europe and the Philippines. However, by the 18th century, declining silver production and economic diversification greatly diminished royal income. In response, the Crown enacted the Bourbon Reforms, a series of edicts that increased taxes and partitioned the Viceroyalty of Peru. The new laws provoked Tupac Amaru II's rebellion and other revolts, all of which were defeated.

In the early 19th century, while most of South America was swept by wars of independence, Peru remained a royalist stronghold. As the elite hesitated between emancipation and loyalty to the Spanish Monarchy, independence was achieved only after the military campaigns of Jose de San Martin and Simon Bolivar. During the early years of the Republic, endemic struggles for power between military leaders caused political instability. National identity was forged during this period, as Bolivarian projects for a Latin American Confederation foundered and a union with Bolivia proved ephemeral. Between the 1840s and 1860s, Peru enjoyed a period of stability under the presidency of Ramon Castilla through increased state revenues from guano exports. However, by the 1870s, these resources had been squandered, the country was heavily indebted, and political in-fighting was again on the rise.

Peru was defeated by Chile in the 1879–1883 War of the Pacific, losing the provinces of Arica and Tarapaca in the treaties of Ancon and Lima. During the Chilean occupation of Lima, Chilean military authorities turned the University of San Marcos and the recently inaugurated Palacio de la Exposition into barracks, raided medical schools and other educational institutions, plundered the contents of the Peruvian National Library and transported thousands of books (including many centuries-old original volumes) along with much capital stock to Santiago de Chile, and carried away a series of monuments and artwork that had adorned the city. Internal struggles after the war were followed by a period of stability under the Civilista Party, which lasted until the onset of the authoritarian regime of Augusto B. Leguia. The Great Depression caused the downfall of Leguia, renewed political turmoil, and the emergence of the American Popular Revolutionary Alliance (APRA). The rivalry between this organization and a coalition of the elite and the military defined Peruvian politics for the following three decades.

In 1968, the Armed Forces, led by General Juan Velasco Alvarado, staged a coup against president Fernando Belaunde. The new regime undertook radical reforms aimed at fostering development but failed to gain widespread support. In 1975, Velasco was forcefully replaced as president by General Francisco Morales Bermudez, who paralyzed reforms and oversaw the re-establishment of democracy.

During the 1980s, Peru faced a considerable external debt, ever-growing inflation, a surge in drug trafficking, and massive political violence. Some 70,000 people died during the conflict between state forces and Maoist Shining Path guerrillas. Under the presidency of Alberto Fujimori (1990–2000), the country started to recover; however, accusations of authoritarianism, corruption, and human rights violations forced his resignation after the controversial 2000 elections. Since the end of the Fujimori regime, Peru has tried to fight corruption while sustaining economic growth; since 2006 the president is Alan Garcia.

Government

Peru is a presidential representative democratic republic with a multi-party system. Under the current constitution, the President is the head of state and government; he or she is elected for five years and cannot seek immediate re-election, he or she must stand down for at least one full constitutional term before re-election. The President designates the Prime Minister and, with his advice, the rest of the Council of Ministers. There is a unicameral Congress with 120 members elected for a five-year term. Bills may be proposed by either the executive or the legislative branch; they become law after being passed by Congress and promulgated by the President. The judiciary is nominally independent, though political intervention into judicial matters has been common throughout history and arguably continues today.

The Peruvian government is directly elected, and voting is compulsory for all citizens aged 18 to 70. General elections held in 2006 ended in a second round victory for presidential candidate Alan Garcia of the Peruvian Aprista Party (52.6% of valid votes) over Ollanta Humala of Union for Peru (47.4%). Congress is currently composed of the Peruvian Aprista Party (36 seats), Peruvian Nationalist Party (23 seats), Union for Peru (19 seats), National Unity (15 seats), the Fujimorista Alliance for the Future (13 seats), the Parliamentary Alliance (9 seats) and the Democratic Special Parliamentary Group (5 seats).

Peruvian foreign relations have been dominated by border conflicts with neighboring countries, most of which were settled during the 20th century. There is still an ongoing dispute with Chile over maritime limits in the Pacific Ocean. Peru is an active member of several regional blocs and one of the founders of the Andean Community of Nations. It is also a participant in international organizations such as the Organization of American States and the United Nations. The Peruvian military is composed of an army, a navy and an air force; its primary mission is to safeguard the independence, sovereignty and territorial integrity of the country. The armed forces are subordinate to the Ministry of Defense and to the President as Commander-in-Chief. Conscription was abolished in 1999 and replaced by voluntary military service.

Regions

Peru is divided into 25 regions and the province of Lima. Each region has an elected government composed of a president and a council, which serves for a four-year term. These governments plan regional development, execute public investment projects, promote economic activities, and manage public property. The province of Lima is administered by a city council.

Regions:

- Amazonas

- Ancash
- Apurimac
- Arequipa
- Ayacucho
- Cajamarca
- Callao
- Cusco
- Huancavelica
- Huanuco
- Ica
- Junin
- La Libertad
- Lambayeque
- Lima
- Loreto
- Madre de Dios
- Moquegua
- Pasco
- Piura
- Puno
- San Martin
- Tacna
- Tumbes
- Ucayali.

Geography

Peru covers 1,285,220 km² (496,226 sq mi). It neighbours Ecuador and Colombia to the north, Brazil to the east, Bolivia to the southeast, Chile to the south, and the Pacific Ocean to the west.

The Andes mountains run parallel to the Pacific Ocean, dividing the country into three geographic regions. The *costa* (coast), to the west, is a narrow plain, largely arid except for valleys created by seasonal rivers. The *sierra* (highlands) is the region of the Andes; it includes the *Altiplano* plateau as well as the highest peak of the country, the 6,768 m (22,205 ft)

Huascaran. The third region is the *selva* (jungle), a wide expanse of flat terrain covered by the Amazon rainforest that extends east. Almost 60% of the country's area is located within this region, (70,000,000 hectares / 172,973,767 acres or 700,000 km/ 270,272 sq mi) giving Peru the fourth largest area of tropical forest in the world after Brazil, Congo and Indonesia.

Most Peruvian rivers originate in the peaks of the Andes and drain into one of three basins. Those that drain toward the Pacific Ocean are steep and short, flowing only intermittently. Tributaries of the Amazon River are longer, have a much larger flow, and are less steep once they exit the *sierra*. Rivers that drain into Lake Titicaca are generally short and have a large flow. Peru's longest rivers are the Ucayali, the Maranon, the Putumayo, the Yavari, the Huallaga, the Urubamba, the Mantaro, and the Amazon.

Peru, unlike other equatorial countries, does not have an exclusively tropical climate; the influence of the Andes and the Humboldt Current cause great climatic diversity within the country. The *costa* has moderate temperatures, low precipitations, and high humidity, except for its warmer, wetter northern reaches. In the *sierra*, rain is frequent during summer, and temperature and humidity diminish with altitude up to the frozen peaks of the Andes. The *selva* is characterized by heavy rainfall and high temperatures, except for its southernmost part, which has cold winters and seasonal rainfall. Because of its varied geography and climate, Peru has a high biodiversity with 21,462 species of plants and animals reported as of 2003; 5,855 of them endemic. The Peruvian government has established several protected areas for their preservation.

Economy

Peru's economy has experienced significant growth in the last 15 years . It is considered an Emerging Market according to the MSCI. Peru had a high Human Development Index score of 0.806, according to a 2008 report. Its 2008 per capita income was US$8,594; by 2009, 34.8% of its total population is poor, including 11.5% that is extremely poor. Historically, the country's economic performance has been tied to exports, which provide hard currency to finance imports and external debt payments. Although exports have provided substantial revenue, self-sustained growth and a more egalitarian distribution of income have proven elusive.

Peruvian economic policy has varied widely over the past decades. The 1968–1975 government of Juan Velasco Alvarado introduced radical reforms, which included agrarian reform, the expropriation of foreign companies, the introduction of an economic planning system, and the creation of a large state-owned sector. These measures failed to achieve their objectives of income redistribution and the end of economic dependence on developed nations.

Despite these adverse results, most reforms were not reversed until the 1990s, when the liberalizing government of Alberto Fujimori ended price controls, protectionism, restrictions on foreign direct investment, and most state ownership of companies. Reforms have permitted sustained economic growth since 1993, except for a slump after the 1997 Asian financial crisis.

Services account for 53% of Peruvian gross domestic product, followed by manufacturing (22.3%), extractive industries (15%), and taxes (9.7%). Recent economic growth has been fueled by macroeconomic stability, improved terms of trade, and rising investment and consumption. Trade is expected to increase further after the implementation of a free trade agreement with the United States signed on April 12, 2006. Peru's main exports are copper, gold, zinc, textiles, and fish meal; its major trade partners are the United States, China, Brazil, and Chile.

Demographics

With about 29 million inhabitants, Peru is the fourth most populous country in South America as of 2007. Its demographic growth rate declined from 2.6% to 1.6% between 1950 and 2000; population is expected to reach approximately 42 million in 2050. As of 2007, 75.9% lived in urban areas and 24.1% in rural areas. Major cities include Lima, home to over 8 million people, Arequipa, Trujillo, Chiclayo, Piura, Iquitos, Cusco, Chimbote, and Huancayo, all of which reported more than 250,000 inhabitants in the 2007 census. In the Amazonian region, there are 16 ethno-linguistic families and more than 65 different ethnic groups. After Brazil and New Guinea, Peru has the largest number of uncontacted tribes in the world.

Peru is a multiethnic country formed by the combination of different groups over five centuries. Amerindians inhabited Peruvian territory for several millennia before Spanish Conquest in the 16th century; their population decreased from an estimated 9 million in the 1520s to around 600,000 in 1620 mainly because of infectious diseases. The laws made to protect the indigenous people are not always respected by the Peruvian government or the companies, such as Perenco, Repsol YPF, and Petrobras, who seek to explore the natural resources of their land. In 1994, Peru signed and ratified the current international law concerning indigenous peoples, the Indigenous and Tribal Peoples Convention, 1989. Spaniards and Africans arrived in large numbers under colonial rule, mixing widely with each other and with indigenous peoples. After independence, there has been a gradual European immigration from England, France, Germany, Italy, and Spain. Chinese arrived in the 1850s as a replacement for slave workers and have since become a major influence in Peruvian society. Other immigrant groups include Arabs and Japanese. Given Peru's high rate of ethnic intermixing, the country's racial structure can be loosely classified as 31%-45% Amerindian, 37% Mestizo (mixed Amerindian and European), 15% European, and 3% African, Japanese, Chinese, and others.

Spanish, the first language of 83.9% of Peruvians aged five and older in 2007, is the primary language of the country. It coexists with several indigenous languages, the most important of which is Quechua, spoken by 13.2% of the population. Other native and foreign languages were spoken at that time by 2.7% and 0.1% of Peruvians, respectively. In the 2007 census, 81.3% of the population over 12 years old described themselves as Catholic, 12.5% as Evangelical, 3.3% as of other denominations, and 2.9% as non-religious.

Education and Health

Literacy was estimated at 92.9% in 2007; this rate is lower in rural areas (80.3%) than in urban areas (96.3%). Primary and secondary education are compulsory and free in public schools. Schools in rural areas are associated with less average knowledge of pupils. There is a correlation of malnutrition and low achievement at school. Income shows a positive correlation with education. In health care, discrimination against indigenous peoples exists. Peru has one of the highest maternal death rates of the Americas. In the 1996-2006 period, infant mortality was at 21 per 1,000 births.

Culture

Peruvian culture is primarily rooted in Amerindian and Spanish traditions, though it has also been influenced by various African, Asian, and European ethnic groups. Peruvian artistic traditions date back to the elaborate pottery, textiles, jewelry, and sculpture of Pre-Inca cultures. The Incas maintained these crafts and made architectural achievements including the construction of Machu Picchu. Baroque dominated colonial art, though modified by native traditions. During this period, most art focused on religious subjects; the numerous churches of the era and the paintings of the Cuzco School are representative. Arts stagnated after independence until the emergence of *Indigenismo* in the early 20th century. Since the 1950s, Peruvian art has been eclectic and shaped by both foreign and local art currents.

Peruvian literature has its roots in the oral traditions of pre-Columbian civilizations. Spaniards introduced writing in the 16th century; colonial literary expression included chronicles and religious literature. After independence, Costumbrism and Romanticism became the most common literary genres, as exemplified in the works of Ricardo Palma. In the early 20th century, the *Indigenismo* movement produced such writers as Ciro Alegría, Jose Maria Arguedas, and Cesar Vallejo. During the second half of the century, Peruvian literature became more widely known because of authors such as Mario Vargas Llosa, a leading member of the Latin American Boom.

Peruvian cuisine is based on Spanish food during the colonial period with various influences from the traditional cuisine of Amerindians. In Lima, there have also been minor influences from African, Chinese, and Japanese immigrants (who make up less than three percent of the population and live mostly in Lima). These cuisines have been heavily adapted due to a lack of authentic ingredients many of which are either commercially unavailable or prohibitively expensive. For example, although there are numerous Chinese restaurants throughout Lima, none of them serve authentic Chinese cuisine. Instead they serve a fusion cuisine known as Chifa. Common dishes include *anticuchos*, *ceviche*, *humitas*, and *pachamanca*. Because of the variety of climates within Peru, a range of plants and animals is available for cooking.

Peruvian music has Andean, Spanish and African roots. In pre-Hispanic times, musical expressions varied widely from region to region; the *quena* and the *tinya* were two common instruments. Spanish conquest brought the introduction of new instruments such as the guitar and the harp, as well as the development of crossbred instruments like the *charango*. African contributions to Peruvian music include its rhythms and the *cajon*, a percussion instrument. Peruvian folk dances include *marinera*, *tondero*, *danza de tijeras*, *huayno* and *diablada*.

7

Bolivia

Bolivia, officially known as the Plurinational State of Bolivia, is a landlocked country in central South America. It is bordered by Brazil to the North and East, Paraguay and Argentina to the South, and Chile and Peru to the West.

Prior to European colonization, the Andean region of Bolivia was a part of the Inca Empire - the largest state in Pre-Columbian America. The Spanish Empire conquered the region in the 16th century. During most of the Spanish colonial period, this territory was called *Upper Peru* or *Charcas* and was under the administration of the Viceroyalty of Peru, which included most of Spain's South American colonies. After declaring independence in 1809, 16 years of war followed before the establishment of the Republic, named for Simon Bolivar, on August 6, 1825. Bolivia has struggled through periods of political instability, dictatorships and economic woes.

Bolivia is a Democratic Republic, divided into nine departments. Its geography is varied from the peaks of the Andes in the West, to the Eastern Lowlands, situated within the Amazon Basin. It is a developing country, with a Medium Human Development Index score, and a poverty level around 60%. Its main economic activities include agriculture, forestry, and fishing, mining and manufacturing goods such as textiles, clothing, refined metals, and refined petroleum. Bolivia is very wealthy in minerals, especially tin.

The Bolivian population, estimated at 10 million, is multiethnic, including Amerindians, Mestizos, Europeans, Asians and Africans. The main language spoken is Spanish, although the Aymara and Quechua languages are also common and all three, as well as 36 other indigenous languages, are official. The large number of different cultures within Bolivia has contributed greatly to a wide diversity in fields such as art, cuisine, literature, and music.

Etymology

The word *Bolivia* is derived from *Bolivar*, the last name of the South American *Libertador* Simon Bolivar. The name came about when Antonio Jose de Sucre was given the option by Bolivar to either keep *Upper Peru* (present-day Bolivia) under the newly formed Republic of Peru, to unite with the United Provinces of Rio de la Plata, or to formally declare its independence from the Viceroyalty of Peru that had dominated most of the region. Sucre opted to create a new nation and, with local support, named it in honour of Simon Bolivar.

However, the original name given to the newly formed country was *Republic of Bolivar*. The name would not change to *Bolivia* until some days later when congressman *Manuel Martin Cruz* proposed: *"If from Romulus comes Rome, then from Bolivar comes Bolivia"* (Spanish: *Si de Romulo Roma, de Bolivar Bolivia*). The name stuck and was approved by the Republic on October 3, 1825.

History

The region that is now known as Bolivia has been constantly occupied for over 2,000 years, when the Aymara arrived in the region. Present-day Aymara associate themselves with an advanced civilization situated at Tiwanaku, in Western Bolivia. The capital city of Tiwanaku dates as early as 1500 BC as a small agriculturally based village.

The community grew to urban proportions between AD 600 and AD 800, becoming an important regional power in the southern Andes. According to early estimates, at its maximum extent, the city covered approximately 6.5 square kilometres, and had between 15,000 – 30,000 inhabitants. However, satellite imaging was used recently to map the extent of fossilized suka kollus across the three primary valleys of Tiwanaku, arriving at population-carrying capacity estimates of anywhere between 285,000 and 1,482,000 people.

Around AD 400, Tiwanaku went from being a locally dominant force to a predatory state. Tiwanaku expanded its reaches into the Yungas and brought its culture and way of life to many other cultures in Peru, Bolivia, and Chile. However, Tiwanaku was not a violent culture in many aspects. In order to expand its reach Tiwanaku became very political creating colonies, trade agreements (which made the other cultures rather dependant), and state cults.

The empire continued to grow with no end in sight. William H. Isbell states that "Tiahuanaco underwent a dramatic transformation between AD 600 and 700 that established new monumental standards for civic architecture and greatly increased the resident population." Tiwanaku continued to absorb cultures rather than eradicate them. Archaeologists have seen a dramatic adoption of Tiwanaku ceramics in the cultures who became part of the Tiwanaku empire. Tiwanaku gained its power through the trade it implemented between all of the cities within its empire.

The elites gained their status by the surplus of food they gained from all of the regions and then by having the ability to redistribute the food among all the people. This is where the control of llama herds became very significant to Tiwanaku. The llama herds were essential for carrying goods back and forth between the centre and the periphery as well as symbolizing the distance between the commoners and the elites. Their power continued to grow in this manner of a surplus of resources until about AD 950. At this time a dramatic shift in climate occurred.

At this point in time there was a significant drop in precipitation for the Titicaca Basin. Some archaeologists even venture to say that a great drought occurred. As the rain became less and less many of the cities further away from Lake Titicaca began to produce fewer crops to give to the elites. As the surplus of food ran out for the elites their power began to fall. The capital city became the last place of production, due to the resiliency of the raised fields, but in the end even the intelligent design of the fields was no match for the weather. Tiwanaku disappeared around AD 1000 because food production, their main source of power, dried up. The land was not inhabited for many years after that.

Between 1438 and 1527, the Incan empire, on a mass expansion, acquired much of what is now western Bolivia. The Incans wouldn't maintain control of the region for long however, as the rapidly expanding Inca Empire was internally weak. As such, the Spanish conquest would be remarkably easy.

Colonial Period

The Spanish conquest began in 1524 and was mostly completed by 1533. The territory now called Bolivia was then known as "Upper Peru" and was under the authority of the Viceroy of Lima. Local government came from the Audiencia de Charcas located in Chuquisaca (La Plata—modern Sucre). Founded in 1545 as a mining town, Potosi soon produced fabulous wealth, becoming the largest city in the New World with a population exceeding 150,000 people.

By the late 16th century Bolivian silver was an important source of revenue for the Spanish Empire. A steady stream of natives served as labour force (the Spanish employed the pre-Columbian draft system called the mita). Upper Peru was bounded to Viceroyalty of the Rio de la Plata in 1776. Tupac Katari led the indigenous rebellion that laid siege to La Paz in March of 1781, during which 20,000 people died. As Spanish royal authority weakened during the Napoleonic wars, sentiment against colonial rule grew.

Independence and Subsequent Wars

The struggle for independence started on the city of Sucre on 1809, and after 16 years of war the Republic was proclaimed on August 6, 1825, named for Simon Bolivar.

In 1836, Bolivia, under the rule of Marshal Andres de Santa Cruz, invaded Peru to reinstall the deposed president, General Luis Jose de Orbegoso. Peru and Bolivia formed the Peru-Bolivian Confederation, with de Santa Cruz as the *Supreme Protector*. Following tension between the Confederation and Chile, Chile declared war on December 28, 1836. Argentina, Chile's ally, declared war on the Confederation on May 9, 1837. The Peruvian-Bolivian forces achieved several major victories during the War of the Confederation: the defeat of the Argentinian expedition and the defeat of the first Chilean expedition on the fields of Paucarpata near the city of Arequipa.

On the same field the Paucarpata Treaty was signed with the unconditional surrender of the Chilean and Peruvian rebel army. The treaty stipulated that Chile withdraw from Peru-Bolivia, return captured Confederate ships, economic relations would be normalized, and the Confederation would pay Peruvian debt to Chile. In Chile public outrage over the treaty forced the government to reject it. Chile organized a second attack on the Confederation and defeated it in the Battle of Yungay. After this defeat, Santa Cruz resigned and went to exile in Ecuador and then Paris, and the Peruvian-Bolivian Confederation was dissolved.

Following the independence of Peru, Peruvian president General Agustin Gamarra invaded Bolivia. The Peruvian army was decisively defeated at the Battle of Ingavi on November 20, 1841, where Gamarra was killed. The Bolivian army under General Jose Ballivian then mounted a counter-offensive managing to capture the Peruvian port of Arica. Later, both sides signed a peace treaty, the Declaration of Independence of Bolivia, in 1842, putting a final end to the war.

Economic Instability and Continued Wars

A period of political and economic instability in the early to mid-19th century weakened Bolivia. Then in the War of the Pacific (1879–83) against Chile, it lost its access to the sea and the adjoining rich *salitre* (saltpeter) fields, together with the port of Antofagasta.

Since its independence, Bolivia has lost over half of its territory to neighboring countries in wars. It also lost the state of Acre, in the Acre War; this region was (known for its production of rubber). Peasants and the Bolivian army fought briefly but after a few victories, and facing the prospect of a total war against Brazil, it was forced to sign the Treaty of Petrópolis in 1903, in which Bolivia lost this rich territory.

In the late 19th century, an increase in the world price of gold brought Bolivia relative prosperity and political stability. During the early 20th century, tin replaced gold as the country's most important source of wealth. A succession of governments controlled by the economic and social elite followed laissez-faire capitalist policies through the first thirty years of the 20th century.

Living conditions of the native people, who constitute most of the population, remained deplorable. With work opportunities limited to primitive conditions in the mines and in large estates having nearly feudal status, they had no access to education, economic opportunity, and political participation. Bolivia's defeat by Paraguay in the Chaco War (1932–35), where Bolivia lost a great part of the Gran Chaco region in dispute, marked a turning-point.

Nationalist Revolutionary Movement

The Revolutionary Nationalist Movement (MNR) emerged as a broadly based party. Denied its victory in the 1951 presidential elections, the MNR led a successful revolution in 1952. Under President Victor Paz Estenssoro, the MNR, having strong popular pressure, introduced universal suffrage into his political platform and carried out a sweeping land-reform promoting rural education and nationalization of the country's largest tin mines.

Twelve years of tumultuous rule left the MNR divided. In 1964, a military junta overthrew President Estenssoro at the outset of his third term. The 1969 death of President René Barrientos Ortuno, a former member of the junta elected president in 1966, led to a succession of weak governments. Alarmed by the rising Popular Assembly and the increase in popularity of The Presidente Juan Jose Torres, the military, the MNR, and others installed Coronel (later General) Hugo Banzer Suarez as president in 1971.

Banzer ruled with MNR support from 1971 to 1974. Then, impatient with schisms in the coalition, he replaced civilians with members of the armed forces and suspended political activities. The economy grew impressively during most of Banzer's presidency, but human rights violations and eventual fiscal crises undercut his support. He was forced to call elections in 1978, and Bolivia again entered a period of political turmoil.

CIA Activities and Execution of Che Guevara

CIA had been active in providing finances and training to the Bolivian military in 1960s. The revolutionary leader Che Guevara was executed by a team of CIA officers and Bolivian Army on 9 October 1967 in Bolivia. CIA reported that Guevara was captured on 8 October as a result of the clash with the Cuban-led guerrillas. He had a wound in his leg, but was otherwise in fair condition. At 1150 hours on 9 October the Second Ranger Battalion received direct orders from Bolivian Army Headquarters in La Paz to kill Guevara. These orders were carried out at 1315 hours the same day with a burst of fire from an M-2 automatic rifle. Felix Rodriguez was a CIA officer on the team with the Bolivian Army that captured and shot Guevara. Rodriquez said that after he received a Bolivian presidential execution order, he told "the soldier who pulled the trigger to aim carefully, to remain consistent with the Bolivian government's story that Che had been killed in action during a clash with the Bolivian army." Rodriguez said the US government had wanted Che in Panama, and "I could have tried to falsify the command to the troops, and got Che to Panama as the US government said they had wanted," said Mr. Rodriguez.

He chose to "let history run its course" as desired by Bolivia.

Military Governments: Garcia Meza and Siles Zuazo

Elections in 1979 and 1981 were inconclusive and marked by fraud. There were coups d'etat, counter-coups, and caretaker governments. In 1980, General Luis Garcia Meza Tejada carried out a ruthless and violent coup d'etat that did not have popular support. He pacified the people by promising to remain in power only for one year. (At the end of the year, he staged a televised rally to claim popular support and announced, "*Bueno, me quedo*," or, "All right; I'll stay [in office]." He was deposed shortly thereafter.) His government was notorious for human-rights-abuses, narcotics-trafficking, and economic mismanagement; during his presidency, the inflation that later crippled the Bolivian economy could already be felt. Later convicted in absentia for various crimes, including murder, Garcia Meza was extradited from Brazil and began serving a 30-year sentence in 1995.

After a military rebellion forced out Meza in 1981, three other military governments in 14 months struggled with Bolivia's growing problems. Unrest forced the military to convoke the Congress elected in 1980 and allow it to choose a new chief executive. In October 1982, Hernan Siles Zuazo again became president, 22 years after the end of his first term of office (1956–60).

Sanchez de Lozada and Banzer: Liberalizing the Economy

Gonzalo Sanchez de Lozada pursued an aggressive economic and social reform agenda. The most dramatic reform was the "capitalization" program, under which investors, typically foreign, acquired 50% ownership and management control of public enterprises, such as the state oil corporation, telecommunications system, airlines, railroads, and electric utilities, in return for agreed upon capital investments.

The reforms and economic restructuring were strongly opposed by certain segments of society, which instigated frequent and sometimes violent protests, particularly in La Paz and the Chapare coca-growing region, from 1994 through 1996. The de Lozada government pursued a policy of offering monetary compensation for voluntary eradication of illegal coca by its growers in the Chapare region. The policy produced little net reduction in coca, and in the mid-1990s Bolivia accounted for about one-third of the world's coca that was being processed into cocaine. The coca leaf has long been part of the Bolivian culture, as indigenous workers have traditionally used the leaf for its properties as a mild stimulant and appetite suppressant.

During this time, the umbrella labour-organization of Bolivia, the Central Obrera Boliviana (COB), became increasingly unable to effectively challenge government policy. A teachers' strike in 1995 was defeated because the COB could not marshal the support of many of its members, including construction and factory workers. The state also used selective martial law to keep the disruptions caused by the teachers to a minimum. The teachers were led by Trotskyites, and were considered to be the most militant union in the COB. Their downfall was a major blow to the COB, which also became mired in internal corruption and infighting in 1996.

In the 1997 elections, General Hugo Banzer, leader of the Nationalist Democratic Action party (ADN) and former dictator (1971–1978), won 22% of the vote, while the MNR candidate won 18%. General Banzer formed a coalition of the ADN, MIR, UCS, and CONDEPA parties, which held a majority of seats in the Bolivian Congress. The Congress elected him as president, and he was inaugurated on August 6, 1997. During the election-campaign, Banzer had promised to suspend the privatization of the state-owned oil-company, YPFB. But this seemed unlikely to happen, considering Bolivia's weak position globally. The Banzer government basically continued the free-market and privatization-policies of its predecessor.

The relatively robust economic growth of the mid-1990s continued until about the third year of its term in office. After that, regional, global and domestic factors contributed to a decline in economic growth.

Financial crises in Argentina and Brazil, lower world prices for export-commodities, and reduced employment in the coca-sector depressed the Bolivian economy. The public also perceived a significant amount of public-sector corruption. These factors contributed to increasing social protests during the second half of Banzer's term.

At the outset of his government, President Banzer launched a policy of using special police-units to physically eradicate the illegal coca of the Chapare region. The policy produced a sudden and dramatic four-year decline in Bolivia's illegal coca-crop, to the point that Bolivia became a relatively small supplier of coca for cocaine. Those left unemployed by coca-eradication streamed into the cities, especially El Alto, the slum-neighborhood of La Paz. The MIR of Jaime Paz Zamora remained a coalition-partner throughout the Banzer government, supporting this policy (called the Dignity Plan).

On August 6, 2001, Banzer resigned from office after being diagnosed with cancer. He died less than a year later. Vice President Jorge Fernando Quiroga Ramirez completed the final year of his term.

In the June 2002 national elections, former President Gonzalo Sanchez de Lozada (MNR) placed first with 22.5% of the vote, followed by coca-advocate and native peasant-leader Evo Morales (Movement Toward Socialism, MAS) with 20.9%. Morales edged out populist candidate Manfred Reyes Villa of the New Republican Force (NFR) by just 700 votes nationwide, earning a spot in the congressional run-off against Sanchez de Lozada on August 4, 2002.

A July agreement between the MNR and the fourth-place MIR, which had again been led in the election by former President Jaime Paz Zamora, virtually ensured the election of Sanchez de Lozada in the congressional run-off, and on August 6 he was sworn in for the second time. The MNR platform featured three overarching objectives: economic reactivation (and job creation), anti-corruption, and social inclusion.

In 2003 the Bolivian gas conflict broke out. On October 12, 2003 the government imposed martial law in El Alto after sixteen people were shot by the police and several dozen wounded in violent clashes which erupted when a caravan of oil trucks escorted by police and soldiers deploying tanks and heavy-caliber machine guns tried to breach a barricade. On 17 October 2003 Evo Morales' supporters from Cochabamba tried to march into Santa Cruz de la Sierra, the largest city of the eastern lowlands where support was strong for the president. They were turned back. Faced with the option of resigning or more bloodshed, Sanchez de Lozada offered his resignation in a letter to an emergency session of Congress. After his resignation was accepted and his vice president, Carlos Mesa, invested, he left on a commercially scheduled flight for the United States.

In March 2004, the new president Carlos Mesa announced that his government would hold a series of rallies around the country, and at its embassies abroad, demanding that Chile return to Bolivia a stretch of seacoast that the country lost in 1884 after the end of the War of the Pacific. Chile has traditionally refused to negotiate on the issue, but Mesa nonetheless made this policy a central point of his administration.

However, the country's internal situation became unfavourable for such political action on the international stage. After a resurgence of gas protests in 2005, Carlos Mesa attempted to resign in January 2005, but his offer was refused by Congress. On March 22, 2005, after weeks of new street protests from organizations accusing Mesa of bowing to U.S. corporate interests, Mesa again offered his resignation to Congress, which was accepted on June 10. The chief justice of the Supreme Court, Edurrdo Rodriguez, was sworn as interim president to succeed the outgoing Carlos Mesa.

Plan de Todos

Mobilizing against neoliberalism as a common enemy, the indigenous population of the Andean region was able to achieve widespread government reform . Bolivia, in particular, was quite successful due to the prominence of an indigenous population and the persistence of reformist policies. In 1993, Gonzalo Sanchez de Lozada ran for presidency in alliance with the Tupac Katari Revolutionary Liberation Movement, which inspired indigenous-sensitive and multicultural-aware policies . Gonzalo Sanchez de Lozada (colloquially known as Goni) was able to shift Bolivian society by selling state firms and constitutionally acknowledging the existence of a multicultural and multiethnic population. Current development has led to a neoliberal citizenship regime in which civil rights are expressed through private property ownership, formal democracy and representation, and an investment in the maintaining of infrastructure.

In the 1990s, Bolivia introduced, the *Plan de Todos*, which led to the decentralization of government, introduction of intercultural bilingual education, implementation of agrarian legislation, and privatization of state owned businesses. The *Plan de Todos* main incentive was to encourage popular participation among the Bolivian people. The law recognizes the existence of Barrios and rural communities as Territorially Based Organizations (TBOs) and has oversight boards known as *romiles de agilancia*, or vigilance committees, that are responsible for overseeing municipal governments and planning projects. The *Plan* formally acknowledged the existence of 311 municipalities, which benefited directly based on the size of their populations. The *Plan de Todos* inspired the development of a market democracy with minimally regulated capitalist economy. The *Plan* explicitly stated that Bolivian citizens would own a minimum of 51% of enterprises; under the Plan, most state owned enterprises (SOEs), besides mines, were sold . This privatization of SOEs led to innovative neoliberal structuring that acknowledged a diverse population within Bolivia .

The Law of Popular Participation gave municipalities the responsibility of maintaining various infrastructures (and offering services): health, education, systems of irrigation, which stripped the responsibility away from the state. The state provides municipalities with twenty percent of federal tax revenue so that each municipality can adequately maintain these infrastructures. The Law also redistributes political power to the local level.

Bolivia under the Morales Administration

The 2005 Bolivian presidential election was held on December 18, 2005. The two main candidates were Juan Evo Morales Ayma of the MAS Party and Jorge Quiroga, leader of the Social and Democratic Power (PODEMOS) Party and former head of the Accion Democratica Nacionalista (ADN) Party.

Morales won the election with 53.7% of the votes, an absolute majority, unusual in Bolivian elections. He was sworn in on January 22, 2006, for a five-year term. Prior to his official inauguration in La Paz, he was inaugurated in an Aymara ritual at the archeological site of Tiwanaku before a crowd of thousands of Aymara people and representatives of leftist movements from across Latin America. Though highly symbolic, this ritual was not historically based and primarily represented native Aymaras — not the main Quechua-speaking population. It is worth noting that since the Spanish conquest in the early 16th century, this region of South America -with a majority native population- has been ruled mostly by descendants of European immigrants.

On May 1, 2006, Morales announced his intent to re-nationalize Bolivian hydrocarbon assets. While stating that the initiative would not be an expropriation, Morales sent Bolivian troops to occupy 56 gas installations simultaneously. Troops were also sent to the two Petrobras-owned refineries in Bolivia, which provide over 90% of Bolivia's refining-capacity. A deadline of 180 days was announced, by which all foreign energy firms were required to sign new contracts giving Bolivia majority ownership and as much as 82% of revenues (the latter for the largest natural gas fields). All such firms signed contracts. Reports from the Bolivian government and the companies involved are contradictory as to plans for future investment.

By far the biggest customer for Bolivian hydrocarbons has been Brazil, which imports two-thirds of Bolivia's natural gas via pipelines operated by the semi-private Petrobras. Since gas can only be exported from landlocked Bolivia via Petrobras' large (and expensive) pipelines, the supplier and customer are strongly linked. Petrobras has announced plans to produce enough natural gas by 2011 to replace that now supplied by Bolivia. Bolivia's position is strengthened by the knowledge that hydrocarbon reserves are more highly valued now than at the times of previous nationalizations, and by the pledged support of President Hugo Chavez of Venezuela.

Fulfilling a campaign promise, Morales opened on August 6, 2006, the Bolivian Constituent Assembly to begin writing a new constitution aimed at giving more power to the indigenous majority. Problems immediately arose when, unable to garner the two-thirds votes needed to include controversial provisions in the constitutional draft, Morales' party announced that only a simple majority would be needed to draft individual articles while two-thirds needed to pass the document in full. Violent protests arose in December 2006 in parts of the country for both two-thirds and departmental autonomy, mostly in the eastern third of the country, where much of the hydrocarbon wealth is located. MAS and its supports believed two-thirds voting rules would give an effective veto for all constitutional changes to the conservative minority.

In August 2007, more conflicts arose in Sucre, as the city demanded the discussion of the seat of government inside the assembly, hoping the executive and legislative branch could return to the city, but assembly and the government said this demand was overwhelmingly impractical and politically undesirable. The conflict turned into violence, and the assembly was moved to a military area in Oruro. Although the main opposition party boycotted the session, a constitutional draft was approved on November 24.

Geography

At 1,098,580 square kilometres (424,160 sq mi), Bolivia is the world's 28th-largest country. Bolivia has been a landlocked nation since 1879, when it lost its coastal department of Litoral to Chile in the War of the Pacific. However, it does have access to the Atlantic via the Paraguay River.

Many ecological zones are represented within Bolivia's territory. The western highlands of the country are situated in the Andes Mountains and include the Bolivian Altiplano. The eastern lowlands include large sections of Amazonian rainforests and the Chaco Plain. The highest peak is Nevado Sajama at 6,542 metres (21,463 ft) located in the Oruro Department. Lake Titicaca is located on the border between Bolivia and Peru. The Salar de Uyuni, the world's largest salt flat, lies in the southwest corner of the country, in the department of Potosi.

Climate

The weather in Bolivia can vary drastically from one climatic zone to another. The summer months in Bolivia are November through March. The weather is typically warmer and wetter during these months. April through October, the winter months, are typically colder and drier.

In the highlands, the weather can be very cold and temperatures frequently go below zero at night, especially on the Altiplano. Snow is common in Potosi during the winter months and sometimes also falls on La Paz and Oruro. In contrast, winter in Sucre, Cochabamba and Tarija on the Cordillera Real is a time of blue skies and comfortable temperatures.

The weather in the rainforest is usually very hot and is often very wet. The drier period of the year is May to October. The section of the rainforest that borders the Cordillera Real of the Andes Mountains is a bit cooler, but still very wet. As altitude declines, the temperature rises.

Geology

The geology of Bolivia compromises a variety of different lithologies as well as tectonic and sedimentary environments. On a synoptic scale geological units coincide with topographical units, to begin the country is divided into a mountainous western area affected by the subduction processes in the Pacific and a eastern lowlands of stable platforms and shields.

Economy

Bolivia has the lowest GDP per capita in South America, despite being rich in natural resources. Bolivia's 2002 gross domestic product (GDP) totaled USD $7.9 billion. Economic growth is about 2.5% per year, and inflation was expected to be between 3% and 4% in 2002 (it was under 2% in 2001).

Bolivia was rated 'Repressed' by the 2010 Index of Economic Freedom.

Bolivia's current lackluster economic situation can be linked to several factors from the past three decades. The first major blow to the Bolivian economy came with a dramatic fall in the price of tin during the early 1980s, which impacted one of Bolivia's main sources of income and one of its major mining-industries. The second major economic blow came at the end of the Cold War in the late 1980s and early 1990s as economic aid was withdrawn by western countries who had previously tried to keep a market-liberal regime in power through financial support.

Since 1985, the government of Bolivia has implemented a far-reaching program of macroeconomic stabilization and structural reform aimed at maintaining price stability, creating conditions for sustained growth, and alleviating scarcity. A major reform of the customs service in recent years has significantly improved transparency in this area. Parallel legislative reforms have locked into place market-liberal policies, especially in the hydrocarbon and telecommunication sectors, that have encouraged private investment. Foreign investors are accorded national treatment, and foreign ownership of companies enjoys virtually no restrictions in Bolivia.

Bolivia has the second largest natural gas reserves in South America. The government has a long-term sales-agreement to sell natural gas to Brazil through 2019. The government held a binding referendum in 2005 on the Hydrocarbon Law.

Bolivia is also estimated to have 50%–70% of the world's lithium. The light metal is used to make high-capacity batteries used in electric cars and such. The spinoff effect of lithium mining could cause Bolivia to become the "Saudi Arabia of the Green World." However, to mine for it would involve disturbing the country's salt flats (called Salar de Uyuni), an important natural feature which boosts tourism in the region. The government doesn't want to destroy this unique natural landscape, to meet the rising world demand for lithium.

In April 2000, Bechtel signed a contract with Hugo Banzer, the former President of Bolivia, to privatize the water supply in Bolivia's third-largest city, Cochabamba. Shortly thereafter, the company tripled the water rates in that city, an action which resulted in protests and rioting among those who could no longer afford clean water. Drawing water from community wells or gathering rainwater was made illegal. Amidst Bolivia's nationwide economic collapse and growing national unrest over the state of the economy, the Bolivian government was forced to withdraw the water contract.

Bolivian exports were $1.3 billion in 2002, from a low of $652 million in 1991. Imports were $1.7 billion in 2002. Bolivian tariffs are a uniformly low 10%, with capital equipment charged only 5%. Bolivia's trade-deficit was $460 million in 2002.

Bolivia's trade with neighboring countries is growing, in part because of several regional preferential trade agreements it has negotiated. Bolivia is a member of the Andean Community of Nations and enjoys nominally free trade with other member countries.

The United States remains Bolivia's largest trading partner. In 2002, the United States exported $283 million of merchandise to Bolivia and imported $162 million.

Agriculture accounts for roughly 15% of Bolivia's GDP. Soybeans are the major cash crop, sold into the Andean Community market.

Bolivia's government remains heavily dependent on foreign assistance to finance development projects. At the end of 2002, the government owed $4.5 billion to its foreign creditors, with $1.6 billion of this amount owed to other governments and most of the balance owed to multilateral development banks. Most payments to other governments have been rescheduled on several occasions since 1987 through the Paris Club mechanism. External creditors have been willing to do this because the Bolivian government has generally achieved the monetary and fiscal targets set by IMF programs since 1987, though economic crisis in recent years have undercut Bolivia's normally good record.

The rescheduling of agreements granted by the Paris Club has allowed the individual creditor countries to apply very soft terms to the rescheduled debt. As a result, some countries have forgiven substantial amounts of Bolivia's bilateral debt. The U.S. government reached an agreement at the Paris Club meeting in December 1995 that reduced by 67% Bolivia's existing debt stock. The Bolivian government continues to pay its debts to the multilateral development banks on time. Bolivia is a beneficiary of the Heavily Indebted Poor Countries (HIPC) and Enhanced HIPC debt relief programs, which by agreement restricts Bolivia's access to new soft loans.

The income from tourism becomes more important. Bolivia's tourist industry has grown gradually since about 1990.

Demographics

Bolivia's ethnic distribution is estimated to be 30% Quechua-speaking and 25% Aymara-speaking Amerindians. The largest of the approximately three dozen native groups are the Quechuas (2.5 million), Aymaras (2 million), then Chiquitano (180,000), and Guarani (125,000). So the full Amerindian population is at 55%; the remaining 30% is mestizo (mixed Amerindian and European), and around 15% are whites.

The white population consists mostly of *criollos*, which in turn consist of families of relatively unmixed Spanish ancestry, descended from the early Spanish colonists. These have formed much of the aristocracy since independence. Other smaller groups within the white population are Germans, who founded the former national airline Lloyd Aereo Boliviano, as well as Italians, Basques, Croats, Russians, Poles and other minorities, many of whose members descend from families that have lived in Bolivia for several generations. Some 40,000 German-speaking Mennonites live in eastern Bolivia.

The Afro Bolivian community numbers more than 0.5% of the population, descended from African slaves that were transported to work in Brazil and then migrated westward into Bolivia. They are mostly concentrated in the Yungas region (Nor Yungas and Sud Yungas provinces) in the department of La Paz. There are also Japanese who are concentrated mostly in Santa Cruz de la Sierra, and Middle Easterners who became prosperous in commerce.

Bolivia is one of the least developed countries in South America. Almost two-thirds of its people, many of whom are subsistence farmers, live in poverty. Population density ranges from less than one person per square kilometre in the southeastern plains to about ten per square kilometre (twenty-five per sq.mi) in the central highlands. As of 2006, the population is increasing about 1.45% per year.

Major cities are La Paz (administrative capital), Sucre (capital), Santa Cruz de la Sierra (largest population), El Alto, Oruro and Cochabamba.

Health

In 2006, life expectancy at birth was 64 for males and 67 for females. A study by UN Development Programme and UNICEF reported that over 230 babies in Bolivia died per day through lack of proper care. The majority of the population has no health insurance. A significant part of the population has no access to healthcare.

Religion

The great majority of Bolivians are Roman Catholic, although Protestant denominations and traditional ethnic Incan religion are expanding rapidly. According to a 2001 survey conducted by the National Statistical Institute, 78% of the population is Roman Catholic, 16% is Protestant and 3% follow other religions of Christian origin. Islam practiced by the descendants of Middle Easterners is almost nonexistent. There is also a small Jewish community that is almost all Ashkenazi in origin. The state has no official religion.

There are colonies of Mennonites in the Santa Cruz Department. Many Native communities interweave pre-Columbian and Christian symbols in their worship. According to 2008-2009 studies by Gallup and NGO the population of Bolivia is 59% Catholic, 15% Incan, 12% atheist and agnostic, 11% Protestant (Evangelical) and 3% Bahai, Buddhist and other affiliations.

Language

Spanish 60.7% (official), Quechua 21.2% (official), Aymara 14.6% (official), foreign languages 2.4%, other 1.2% (2001 census according to CIA Factbook). According to Instituto Nacional de Estadistica de Bolivia 28.1% of the population of Bolivia spoke a native language as a first language in 2007. Approximately 90% of the children attend primary-school but often for a year or less. Until the 2001 Census the literacy rate was low in many rural areas, but, according to the CIA, the literacy rate was 87% nationwide, similar to Brazil's but below the South American average. Nevertheless in 2008 after the campaign "Yes I can", Bolivia was declared illiteracy-free under the UNESCO standards.

Politics

Bolivia has been governed by democratically elected governments since 1982, when a long string of military coups came to an end. The country had a total of 193 coups d'etat from independence until 1981, thereby averaging a change of government once every ten months. Presidents Hernan Siles Zuazo (1982–85) and Victor Paz Estenssoro (1985–89) began a tradition of ceding power peacefully.

The 1967 constitution, amended in 1994, provides for balanced executive, legislative, and judicial powers. The traditionally strong executive branch tends to overshadow the Congress, whose role is generally limited to debating and approving legislation initiated by the executive. The judiciary, consisting of the Supreme Court and departmental and lower courts, has long been riddled with corruption and inefficiency. Through revisions to the constitution in 1994, and subsequent laws, the government has initiated potentially far-reaching reforms in the judicial system and processes.

Bolivia's nine departments received greater autonomy under the Administrative Decentralization law of 1995. Departmental autonomy further increased with the first popular elections for departmental governors (*prefectos*) on 18 December 2005, after long protests by pro-autonomy leader department of Santa Cruz.

Bolivian cities and towns are governed by directly elected mayors and councils. Municipal elections were held on 5 December 2004, with councils elected to five year terms. The Popular Participation Law of April 1994, which distributes a significant portion of national revenues to municipalities for discretionary use, has enabled previously neglected communities to make striking improvements in their facilities and services.

The departments of Tarija, Beni, Pando and Santa Cruz are sometimes known as the "half moon" because of the crescent shape of the departments when looked at together in the east of the country. They also have in common conservative politics and rich fossil fuel deposits.

Legislative Branch

Bolivia's government is a republic. The *Asamblea Plurinacional* (National Congress) has two chambers. The *Camara de Diputados* (Chamber of Deputies) has 130 members elected to five year terms, seventy from single-member districts (*circumscriptions*) and sixty by proportional representation. The *Camara de Senadores* (Chamber of Senators) has twenty-seven members (three per department), elected to five year terms.

Administrative Divisions

Bolivia is divided into nine departments (*departements*); capitals in parentheses:

- Beni (Trinidad)
- Chuquisaca (Sucre)
- Cochabamba (Cochabamba)
- La Paz (La Paz)
- Oruro (Oruro)
- Pando (Cobija)
- Potosi (Potosi)
- Santa Cruz (Santa Cruz de la Sierra)
- Tarija (Tarija).

Additionally, the departments are further divided into 100 provinces (*provincias*), and the provinces are each divided into municipalities (*municipios*) and cantons (*cantones*), which handle local affairs.

Military

The Bolivian military comprises three branches: an Army, Navy (despite being landlocked) and Air Force. The legal age for voluntary admissions is 18; however, when the numbers are small the government recruits anyone as young as 14. It is estimated that 20% of the Bolivian army is between the ages 14 and 16 while another 20% is from 16 to 18. The tour of duty is generally 12 months. The Bolivian government annually spends $130 million on defense.

TAB (Transportes Aereos Bolivianos)

Although a civil transport airline, TAB - Transportes Aereos Bolivianos, was created as a subsidiary company of the FAB in 1977. It is subordinate to the Air Transport Management (Gerencia de Transportes Aereos) and is headed by an FAB general. TAB, a charter heavy cargo airline, links Bolivia with most countries of the Western Hemisphere; its inventory included a fleet of Hercules C130 aircraft. TAB's Base of operations was headquartered at El Alto, adjacent to La Paz's El Alto International Airport. TAB also flew to Miami and Houston, with stops in Panama.

Culture

Bolivian culture has been heavily influenced by the Quechua, the Aymara, as well as by the popular cultures of Latin America as a whole.

The cultural development is divided into three distinct periods: precolumbian, colonial, and republican. Important archaeological ruins, gold and silver ornaments, stone monuments, ceramics, and weavings remain from several important pre-Columbian cultures. Major ruins include Tiwanaku, El Fuerte de Samaipata, Incallajta, and Iskanawaya. The country abounds in other sites that are difficult to reach and have seen little archaeological exploration.

The Spanish brought their own tradition of religious art which, in the hands of local native and mestizo builders and artisans, developed into a rich and distinctive style of architecture, painting, and sculpture known as "Mestizo Baroque". The colonial period produced not only the paintings of Perez de Holguín, Flores, Bitti, and others but also the works of skilled but unknown stonecutters, woodcarvers, goldsmiths, and silversmiths. An important body of Native Baroque religious music of the colonial period was recovered in recent years and has been performed internationally to wide acclaim since 1994.

Bolivian artists of stature in the twentieth century include Guzman de Rojas, Arturo Borda, Maria Luisa Pacheco, Roberto Mamani Mamani, Alejandro Mario Yllanes, and Marina Nunez del Prado. Bolivia has a rich folklore. Its regional folk music is distinctive and varied. The "devil dances" at the annual carnival of Oruro are one of the great folkloric events of South America, as is the lesser known carnival at Tarabuco. The best known of the various festivals found in the country is the "Carnaval de Oruro", which was among the first 19 "Masterpieces of the Oral and Intangible Heritage of Humanity," as proclaimed by the UNESCO in May 2001.

Entertainment includes football, which is the national sport, as well as table football, which is played on street corners by both children and adults.

Education

Under UNESCO standards, Bolivia has been declared free of illiteracy in 2008.

Bolivia has a wide variety of public and private universities. Among them: Universidad San Francisco Xavier de Chuquisaca USFX – Sucre, founded in 1624; Universidad Mayor de San Andres UMSA – La Paz, founded in 1830; Universidad Mayor de San Simon UMSS – Cochabamba, founded in 1832; Universidad Autonoma Gabriel Rene Moreno UAGRM – Santa Cruz de la Sierra, founded in 1880; Universidad Tecnica de Oruro UTO – Oruro, founded in 1892; Universidad Autonoma Tomas Frias UATF – Potosi, founded in 1892; Universidad Juan Misael Saracho UJMS – Tarija, founded in 1946; Universidad Catolica Boliviana San Pablo UCB, founded in 1966; Universidad Tecnica del Beni UTB – Trinidad, founded in 1967; Universidad Nur NUR, founded in 1982; Universidad Privada de Santa Cruz de la Sierra UPSC – Santa Cruz de la Sierra, founded in 1984; Universidad Nacional Siglo XX UNSXX – Llallagua, founded in 1986; Universidad del Valle UNIVALLE -Cochabamba, founded in 1988; Universidad Privada Boliviana UPB, founded in 1993; Universidad Privada Franz Tamayo UPFT, founded in 1993 and Universidad Amazonica de Pando UAP – Cobija, founded in 1993.

For the first time in bolivian history, three indigenous universities were created: Universidad Aymara Tupac Katari UATK – La Paz, founded in 2009; Universidad Quechua Casmiro Huanca UQCH – Cochabamba, founded in 2009 and Universidad Boliviana Guarani y Pueblos de Tierras Bajas UGPTB – Chuquisaca, founded in 2009.

8

Chile

Chile, officially the Republic of Chile, is a country in South America occupying a long, narrow coastal strip between the Andes mountains to the east and the Pacific Ocean to the west. It borders Peru to the north, Bolivia to the northeast, Argentina to the east, and the Drake Passage in the far south. With Ecuador, it is one of two countries in South America which do not border Brazil. The Pacific coastline of Chile is 6,435 kilometres. Chilean territory includes the Pacific islands of Juan Fernandez, Salas y Gómez, Desventuradas and Easter Island. Chile also claims about 1,250,000 square kilometres (480,000 sq mi) of Antarctica, although all claims are suspended under the Antarctic Treaty.

The shape of Chile is a distinctive, multi-coloured ribbon of land 4,300 kilometres (2,700 mi) long and on average 175 kilometres (109 mi) wide. Its climate varies, ranging from the world's driest desert-the Atacama-in the north, through a Mediterranean climate in the centre, to a rainy temperate climate in the south. The northern desert contains great mineral wealth, principally copper. The relatively small central area dominates in terms of population and agricultural resources, and is the cultural and political center from which Chile expanded in the late 19th century, when it incorporated its northern and southern regions. Southern Chile is rich in forests and grazing lands and features a string of volcanoes and lakes. The southern coast is a labyrinth of fjords, inlets, canals, twisting peninsulas, and islands. Prior to arrival of the Spanish in the 16th century, northern Chile was under Inca rule while the indigenous Mapuche inhabited central and southern Chile. Chile declared its independence from Spain on February 12, 1818. In the War of the Pacific (1879–83), Chile defeated Peru and Bolivia and won its current northern territory. It was not until the 1880s that the Mapuche were completely subjugated. Although relatively free of the coups and arbitrary governments that blighted South America, Chile endured a 17-year military dictatorship (1973–1990) that left more than 3,000 people dead or missing.

Today, Chile is one of South America's most stable and prosperous nations. It leads Latin American nations in human development, competitiveness, quality of life, political stability, globalization, economic freedom, low perception of corruption and comparatively low poverty rates. It also ranks high regionally in freedom of the press and democratic development. However, it has a high income inequality, as measured by the Gini index. In May 2010 Chile became the first South American country to join the OECD. Chile is also a founding member of both the United Nations and the Union of South American Nations.

Etymology

There are various theories about the origin of the word *Chile.* According to a theory by 17th century Spanish chronicler Diego de Rosales, the Incas of Peru called the valley of the Aconcagua "Chili" by corruption of the name of a Picunche tribal chief ("cacique") called *Tili,* who ruled the area at the time of the Incan conquest in the 15th century. Another theory points to the similarity of the valley of the Aconcagua with that of the Casma Valley in Peru, where there was a town and valley named *Chili.*

Other theories say Chile may derive its name from the native Mapuche word *chilli,* which may mean "where the land ends," "the deepest point of the Earth," or "sea gulls;" or from the Quechua *chin,* "cold", or the Aymara *tchili,* meaning "snow". Another meaning attributed to *chilli* is the onomatopoeic *cheele-cheele*-the Mapuche imitation of a bird call. The Spanish conquistadors heard about this name from the Incas, and the few survivors of Diego de Almagro's first Spanish expedition south from Peru in 1535-36 called themselves the "men of

Chilli." Ultimately, Almagro is credited with the universalization of the name *Chile*, after naming the Mapocho valley as such.

History

About 10,000 years ago, migrating Native Americans settled in fertile valleys and coastal areas of what is present-day Chile. Example settlement sites from the very early human habitation are Cueva del Milodon and the Pali Aike Crater's lava tube. The Incas briefly extended their empire into what is now northern Chile, but the Mapuche (or Araucanians as they were known by the Spaniards) successfully resisted many attempts by the Inca Empire to subjugate them, despite their lack of state organization. They fought against the Sapa Inca Tupac Yupanqui and his army. The result of the bloody three-day confrontation known as the Battle of the Maule was that the Inca conquest of the territories of Chile ended at the Maule river.

In 1520, while attempting to circumnavigate the earth, Ferdinand Magellan discovered the southern passage now named after him, the Strait of Magellan. The next Europeans to reach Chile were Diego de Almagro and his band of Spanish conquistadors, who came from Peru in 1535 seeking gold. The Spanish encountered various cultures that supported themselves principally through slash-and-burn agriculture and hunting. The conquest of Chile began in earnest in 1540 and was carried out by Pedro de Valdivia, one of Francisco Pizarro's lieutenants, who founded the city of Santiago on February 12, 1541. Although the Spanish did not find the extensive gold and silver they sought, they recognized the agricultural potential of Chile's central valley, and Chile became part of the Viceroyalty of Peru.

Conquest of the land took place gradually, and the Europeans suffered repeated setbacks at the hands of the local population. A massive Mapuche insurrection that began in 1553 resulted in Valdivia's death and the destruction of many of the colony's principal settlements. Subsequent major insurrections took place in 1598 and in 1655. Each time the Mapuche and other native groups revolted, the southern border of the colony was driven northward. The abolition of slavery by the Spanish crown in 1683 was done in recognition that enslaving the Mapuche intensified resistance rather than cowing them into submission. Despite the royal prohibitions relations remained strained from continual colonialist interference.

Cut off to the north by desert, to the south by the Mapuche, to the east by the Andes Mountains, and to the west by the ocean, Chile became one of the most centralized, homogeneous colonies in Spanish America. Serving as a sort of frontier garrison, the colony found itself with the mission of forestalling encroachment by both Araucanians and Spain's European enemies, especially the British and the Dutch. In addition to the Araucanians, buccaneers and English adventurers menaced the colony, as was shown by Sir Francis Drake's 1578 raid on Valparaiso, the colony's principal port. Because Chile hosted one of the largest standing armies in the Americas, it was one of the most militarized of the Spanish possessions, as well as a drain on the treasury of the Viceroyalty of Peru. By the end of the colonial period, the population reached an estimated 500,000 (not including unsubjugated Indians); approximately 300,000 were mestizos and about 150,000 were Criollos (European or European descent).

The first general census was performed by the government of Agustin de Jauregui between 1777 and 1778. The census indicated that the population was 259,646 inhabitants and was composed of 73.5% European descent, 7.9% mestizos, 8.6% Indians and 9.8% blacks. In 1784, *Francisco Hurtado*, Governor of the province of Chiloe, conducted a population census of Chiloe whereby the population was 26,703 inhabitants, of which 64.4% were whites and 33.5% natives.

Finally, in 1812, the Diocese of Concepcion made a census of population, south of the Maule river, but not including the indigenous population (estimated at 8,000 people), nor the inhabitants of the province of Chiloe, which gave indicated a population of 210,567, of which 86.1% were Spanish and whites, 10% Indians and 3.7% of mestizos, blacks and mulattos.

The usurpation of the Spanish throne by Napoleon's brother Joseph in 1808 precipitated the drive by the colony for independence from Spain. A national junta in the name of Ferdinand - heir to the deposed king - was formed on September 18, 1810. The Government Junta of Chile proclaimed Chile an autonomous republic within the Spanish monarchy. A movement for total independence soon won a wide following. Spanish attempts to re-impose arbitrary rule during what was called the Reconquista led to a prolonged struggle.

Intermittent warfare continued until 1817, when an army with Bernardo O'Higgins, Chile's most renowned patriot, and led by Jose de San Martin, hero of the Argentine War of Independence, crossed the Andes into Chile and defeated the royalists. On February 12, 1818, Chile was proclaimed an independent republic under O'Higgins' leadership. The political revolt brought little social change, however, and 19th century Chilean society preserved the essence of the stratified colonial social structure, which was greatly influenced by family politics and the Roman Catholic Church. A strong presidency eventually emerged, but wealthy landowners remained powerful.

Toward the end of the nineteenth century, the government in Santiago consolidated its position in the south by ruthlessly suppressing the Mapuche during the Occupation of Araucania. In 1881, it signed a treaty with Argentina confirming Chilean sovereignty over the Strait of Magellan. As a result of the War of the Pacific with Peru and Bolivia (1879–83), Chile expanded its territory northward by almost one-third, eliminating Bolivia's access to the Pacific, and acquired valuable nitrate deposits, the exploitation of which led to an era of national affluence.

The Chilean Civil War in 1891 brought about a redistribution of power between the President and Congress, and Chile established a parliamentary style democracy. However, the Civil War had also been a contest between those who favoured the development of local industries and powerful Chilean banking interests, particularly the House of Edwards who had strong ties to foreign investors.

20th Century

The Chilean economy partially degenerated into a system protecting the interests of a ruling oligarchy. By the 1920s, the emerging middle and working classes were powerful enough to elect a reformist president, Arturo Alessandri Palma, whose program was frustrated by a conservative congress. In the 1920s, Marxist groups with strong popular support arose. A military coup led by General Luis Altamirano in 1924 set off a period of great political instability that lasted until 1932. The longest lasting of the ten governments between those years was that of General Carlos Ibanez del Campo, who briefly held power in 1925 and then again between 1927 and 1931 in what was a de facto dictatorship, although not really comparable in harshness or corruption to the type of military dictatorship that has often bedevilled the rest of Latin America and certainly not comparable to the violent and repressive regime of Augusto Pinochet decades later.

By relinquishing power to a democratically elected successor, Ibanez del Campo retained the respect of a large enough segment of the population to remain a viable politician for more than thirty years, in spite of the vague and shifting nature of his ideology. When constitutional rule was restored in 1932, a strong middle-class party, the Radicals, emerged. It became the key force in coalition governments for the next 20 years. During the period of Radical Party dominance (1932–52), the state increased its role in the economy. In 1952, voters returned Ibanez del Campo to office for another six years. Jorge Alessandri succeeded Ibanez del Campo in 1958, bringing Chilean conservatism back into power democratically for another term.

The 1964 presidential election of Christian Democrat Eduardo Frei Montalva by an absolute majority initiated a period of major reform. Under the slogan "Revolution in Liberty", the Frei administration embarked on far-reaching social and economic programs, particularly in education, housing, and agrarian reform, including rural unionization of agricultural workers. By 1967, however, Frei encountered increasing opposition from leftists, who charged that his reforms were inadequate, and from conservatives, who found them excessive. At the end of his term, Frei had not fully achieved his party's ambitious goals.

In the 1970 election, Senator Salvador Allende a physician and member of the Socialist Party of Chile, who headed "Popular Unity", reached a partial majority in a plurality of votes in a three-way contest, followed by candidates Radomiro Tomic for the Christian Democrat Party and Jorge Alessandri for the Conservative Party. Allende was not elected with an absolute majority, receiving fewer than 355 votes. It became a war of classes, motivated by the central government. Allende was a physician and member of the Socialist Party of Chile, who headed the "Popular Unity" (UP or "Unidad Popular") coalition of the Socialist, Communist, Radical, and Social-Democratic Parties, along with dissident Christian Democrats, the Popular Unitary Action Movement (MAPU), and the Independent Popular Action. Despite pressure from the United States government, the Chilean Congress conducted a runoff vote between the leading candidates, Allende and former president Jorge Alessandri and keeping with tradition, chose Allende by a vote of 153 to 35. Frei refused to form an alliance with Alessandri to oppose Allende, on the grounds that the Christian Democrats were a workers party and could not make common cause with the right-wing.

An economic depression that began in 1972 bottomed out in 1975, exacerbated by capital flight, plummeting private investment, and withdrawal of bank deposits in response to Allende's socialist program. Production fell and unemployment rose. Allende adopted measures including price freezes, wage increases, and tax reforms, to increase consumer spending and redistribute income downward. Joint public-private public works projects helped reduce unemployment. Much of the banking sector was nationalized. Many enterprises within the copper, coal, iron, nitrate, and steel industries were expropriated, nationalized, or subjected to state intervention. Industrial output increased sharply and unemployment fell during the Allende administration's first year.

Allende's program included advancement of workers' interests, replacing the judicial system with "socialist legality", nationalization of banks and forcing others to bankruptcy, and strengthening "popular militias" known as MIR. Started under former President Frei, the Popular Unity platform also called for nationalization of Chile's major copper mines in the form of a constitutional amendment. The measure was passed unanimously by Congress. As a result, the Richard Nixon administration organized and inserted secret operatives in Chile, in order to quickly destabilize Allende's government. In addition, American financial pressure restricted international economic credit to Chile. The economic problems were also exacerbated by Allende's public spending which was financed mostly by printing money and poor credit ratings given by commercial banks. Simultaneously, opposition media, politicians, business guilds and other organizations, helped to accelerate a campaign of domestic political and economical destabilization, some of which was helped by the United States. By early 1973, inflation was out of control. The crippled economy was further battered by prolonged and sometimes simultaneous strikes by physicians, teachers, students, truck owners, copper workers, and the small business class. On 26 May 1973, Chile's Supreme Court, which was opposed to Allende's government, unanimously denounced the Allende *disruption of the legality of the nation.* Although, illegal under the Chilean constitution, the court supported and strengthened Pinochet's seizure of power.

Finally, a military coup overthrew Allende on September 11, 1973. As the armed forces bombarded the presidential palace of (Palacio de La Moneda), Allende reportedly had committed suicide. A military junta, led by General Augusto Pinochet Ugarte, took over control of the country. The first years of the regime were marked by human rights violations. On October 1973, at least 72 people were murdered by the Caravan of Death. According to the Rettig Report and Valech Commission, at least 2,115 were killed, and at least 27,265 were tortured (including 88 children younger than 12 years old). A new Constitution was approved by a controversial plebiscite on September 11, 1980, and General Pinochet became president of the republic for an 8-year term. After Pinochet obtained rule of the country, several hundred committed Chilean revolutionaries joined the Sandinista army in Nicaragua, guerrilla forces in Argentina or training camps in Cuba, Eastern Europe and Northern Africa.

In the late 1980s, the government gradually permitted greater freedom of assembly, speech, and association, to include trade union and political activity. The government launched market-oriented reforms, which have continued ever since. Chile moved toward a free market economy that saw an increase in domestic and foreign private investment, although the copper industry and other important mineral resources were not opened for competition. In a plebiscite on October 5, 1988, General Pinochet was denied a second 8-year term as president (56% against 44%). Chileans elected a new president and the majority of members of a two-chamber congress on December 14, 1989.

Christian Democrat Patricio Aylwin, the candidate of a coalition of 17 political parties called the *Cancerization*, received an absolute majority of votes (55%). President Aylwin served from 1990 to 1994, in what was considered a transition period.

In December 1993, Christian Democrat Eduardo Frei Ruiz-Tagle, the son of previous president Eduardo Frei Montalva, led the Cancerization coalition to victory with an absolute majority of votes (58%).

21st Century

Frei Ruiz-Tagle was succeeded in 2000 by Socialist Ricardo Lagos, who won the presidency in an unprecedented runoff election against Joaquin Lavin of the rightist Alliance for Chile. In January 2006, Chileans elected their first female president, Michelle Bachelet Jeria, of the Socialist Party, defeating Sebastian Pinera, of the National Renewal party, extending the *Cancerization* government for another four years. In January 2010, Chileans elected Sebastian Pinera, of the National Renewal party of the centre-right Coalition for Change, as the first rightist President of Chile during the Chilean presidential election of 2009-2010, defeating former President Eduardo Frei Ruiz-Tagle of the *Cancerization*, for a four-year term succeeding Michelle Bachelet.

On February 27, 2010, Chile was struck by an 8.8 M_W earthquake, one of the largest ever recorded in the world. As many as 500 people died; hundreds of thousands of buildings were damaged. The earthquake was also followed by multiple aftershocks. Initial damage estimates were in the range of 15-30 billion USD, around 10-15% of Chile real gross domestic product. On March 11, 2010 the U.S. Geological Survey reported that a 6.9-magnitude quake hit Chile south of the capital, Santiago.

Geography

A long and narrow coastal Southern Cone country on the west side of the Andes Mountains, Chile stretches over 4,630 kilometres (2,880 mi) north to south, but only 430 kilometres (265 mi) at its widest point east to west. This encompasses a remarkable variety of landscapes. It contains 756,950 square kilometres (292,260 sq mi) of land area. It is situated within the Pacific Ring of Fire. The northern Atacama Desert contains great mineral wealth, primarily copper and nitrates. The relatively small Central Valley, which includes Santiago, dominates the country in terms of population and agricultural resources.

This area also is the historical center from which Chile expanded in the late nineteenth century, when it integrated the northern and southern regions. Southern Chile is rich in forests, grazing lands, and features a string of volcanoes and lakes. The southern coast is a labyrinth of fjords, inlets, canals, twisting peninsulas, and islands. The Andes Mountains are located on the eastern border. Chile is the longest north-south country in the world, and also claims 1,250,000 km^2 (480,000 sq mi) of Antarctica as part of its territory. However, this latter claim is suspended under the terms of the Antarctic Treaty, of which Chile is a signatory.

Chile controls Easter Island and Sala y Gómez Island, the easternmost islands of Polynesia, which it incorporated to its territory in 1888, and Robinson Crusoe Island, more than 600 kilometres (370 mi) from the mainland, in the Juan Fernandez Islands. Easter Island is today a province of Chile. Also controlled but only temporally inhabited (by some local fishermen) are the small islands of Sala y Gómez, San Ambrosio and San Felix. These islands are notable because they extend Chile's claim to territorial waters out from its coast into the Pacific.

Climate

The climate of Chile comprises a wide range of weather conditions across a large geographic scale, extending across 38 degrees in latitude, making generalisations difficult. According to the Koppen system, Chile within its borders hosts at least seven major climatic subtypes, ranging from desert in the north, to alpine tundra and glaciers in the east and south east, humid subtropical in Easter Island, Oceanic in the south and Mediterranean climate in central Chile.

There are four seasons in most of the country: summer (December to February), autumn (March to May), winter (June to August), and spring (September to November).

Flora and Fauna

Chile's botanical zones conform to the topographic and climatic regions. The northernmost coastal and central region is largely barren of vegetation, approaching the most closely an absolute desert in the world. On the slopes of the Andes, besides the scattered tola desert brush, grasses are found. The central valley is characterized by several species of cactus, the hard espinos, the Chilean pine, and the Copihue, a red bell-shaped flower that is Chile's national flower.

In southern Chile, south of the Bio-Bio River, the heavy precipitation has produced dense forests of laurels, magnolias, and various species of conifers and beeches, which become smaller and more stunted to the south. The cold temperatures and winds of the extreme south preclude heavy forestation. Grassland is found in Atlantic Chile (in Patagonia). Much of the Chilean flora is distinct from that of neighboring Argentina, indicating that the Andean barrier existed during its formation. Chilean species include the monkey-puzzle tree, part of the pine-like *Araucaria* genus entered in New Caledonia, and southern beeches.

Chile's geographical isolation also has restricted the immigration of faunal life, so that only a few of the many distinctive Latin American animals are found. Among the larger mammals are the puma or cougar, the llama-like Guanaco, the Andean Wolf, and the fox-like chilla.

In the forest region, several types of marsupials and a small deer known as the pudu are found. There are many species of small birds, but most of the larger common Latin American types are absent. Few freshwater fish are native, but North American trout have been successfully introduced into the Andean lakes. Owing to the vicinity of the Humboldt Current, ocean waters abound with fish and other forms of marine life, which in turn support a rich variety of waterfowl, including different penguins. Whales are abundant, and some six species of seals are found in the area.

Time zones

Because of the distance between the mainland and Easter Island, Chile uses 4 different UTC offsets:

- The mainland uses UTC-4, and in summer as daylight saving time UTC-3.
- Easter Island uses UTC-6, and in summer as daylight saving time UTC-5.

Economy

Chile has a dynamic market-oriented economy characterized by a high level of foreign trade. During the early 1990s, Chile's reputation as a role model for economic reform was strengthened when the democratic government of Patricio Aylwin - which took over from the military in 1990 - deepened the economic reform initiated by the military government. Growth in real GDP averaged 8% during the period 1991-1997, but fell to half that level in 1998 because of tight monetary policies implemented to keep the current account deficit in check and because of lower export earnings - the latter a product of the Asian financial crisis. Chile's economy has since recovered and has seen growth rates of 5-7% over the past several years. In 2006, Chile became the country with the highest nominal GDP per capita in Latin America.

After a decade of impressive growth rates, Chile began to experience a moderate economic downturn in 1999, brought on by unfavourable global economic conditions related to the Asian financial crisis, which began in 1997. The economy remained sluggish until 2003, when it began to show clear signs of recovery, achieving 4.0% real GDP growth. The Chilean economy finished 2004 with growth of 6.0%. Real GDP growth reached 5.7% in 2005 before falling back to 4.0% growth in 2006. GDP expanded 5.1% in 2007.

Sound economic policies, maintained consistently since the 1980s, have contributed to steady growth and reduced poverty rates by over half. The 1973-90 military government sold many state-owned companies, and the three democratic governments since 1990 have continued privatization, though at a slower pace. The government's role in the economy is mostly limited to regulation, although the state continues to operate copper giant CODELCO and a few other enterprises (there is one state-run bank). Chile is strongly committed to free trade and has welcomed large amounts of foreign investment. Chile has signed free trade agreements (FTAs) with a whole network of countries, including an FTA with the United States, which was signed in 2003 and implemented in January 2004.

Over the last several years, Chile has signed FTAs with the European Union, South Korea, New Zealand, Singapore, Brunei, China, and Japan. It reached a partial trade agreement with India in 2005 and began negotiations for a full-fledged FTA with India in 2006. Chile conducted trade negotiations in 2007 with Australia, Malaysia, and Thailand, as well as with China to expand an existing agreement beyond just trade in goods. Chile concluded FTA negotiations with Australia and the expanded agreement with China in 2008. The members of the P4 (Chile, Singapore, New Zealand, and Brunei) also plan to conclude a chapter on finance and investment in 2008. The economic international organization the OECD agreed to invite Chile to be among four countries to open discussions in becoming an official member. It was invited to join the organization in December 2009, and accepted in January 2010.

High domestic savings and investment rates helped propel Chile's economy to average growth rates of 8% during the 1990s. The privatized national pension system (AFP) has encouraged domestic investment and contributed to an estimated total domestic savings rate of approximately 21% of GDP.

Unemployment hovered in the 8%-10% range after the start of the economic slowdown in 1999, above the 7% average for the 1990s. Unemployment finally dipped to 7.8% for 2006, and has kept falling in 2007, averaging 6.8% monthly (up to August). Wages have risen faster than inflation as a result of higher productivity, boosting national living standards. The percentage of Chileans with household incomes below the poverty line-defined as twice the cost of satisfying a person's minimal nutritional needs-fell from 45.1% in 1987 to 13.7% in 2006, according to government polls. Critics in Chile, however, argue true poverty figures are considerably higher than those officially published, because the government constructs the poverty line based on an outdated 1987 household consumption poll, instead of the more recent ones from 1997 or 2007. According to these critics, using data from the 1997 poll, the poverty rate rises to 29%. Using the relative yardstick favoured in many European countries, 27% of Chileans would be poor, according to Juan Carlos Feres of the ECLAC. Chile's independent Central Bank pursues an inflation target of between 2% and 4%. Inflation has not exceeded 5% since 1998. Chile registered an inflation rate of 3.2% in 2006. The Chilean peso's rapid appreciation against the U.S. dollar in recent years has helped dampen inflation. Most wage settlements and loans are indexed, reducing inflation's volatility. Under the compulsory private pension system, most formal sector employees pay 10% of their salaries into privately managed funds.

Total foreign direct investment (FDI) was only $3.4 billion in 2006, up 52% from a poor performance in 2005. However, 80% of FDI continues to go to only four sectors: electricity, gas, water and mining. Much of the jump in FDI in 2006 was also the result of acquisitions and mergers and has done little to create new employment in Chile. The Chilean Government has formed a Council on Innovation and Competition, which is tasked with identifying new sectors and industries to promote. It is hoped that this, combined with some tax reforms to encourage domestic and foreign investment in research and development, will bring in additional FDI and to new parts of the economy. As of 2006, Chile invested only 0.6% of its annual GDP in research and development (R&D). Even then, two-thirds of that was government spending. Beyond its general economic and political stability, the government also has encouraged the use of Chile as an "investment platform" for multinational corporations planning to operate in the region, but this will have limited value given the developing business climate in Chile itself. Chile's approach to foreign direct investment is codified in the country's Foreign Investment Law, which gives foreign investors the same treatment as Chileans. Registration is reported to be simple and transparent, and foreign investors are guaranteed access to the official foreign exchange market to repatriate their profits and capital.

Faced with an international economic downturn the government announced a $4 billion economic stimulus plan to spur employment and growth despite the global financial crisis, aiming for an expansion of between 2 percent and 3 percent of GDP for 2009. Nonetheless, economic analysts differ from the government estimates and forecast economic growth at a median of 1.5 percent.

Foreign Trade

2006 was a record year for Chilean trade. Total trade registered a 31% increase over 2005. During 2006, exports of goods and services totaled US $58 billion, an increase of 41%. This figure was somewhat distorted by the skyrocketing price of copper. In 2006, copper exports reached a historical high of US $33.3 billion. Imports totaled US $35 billion, an increase of 17% compared to the previous year. Chile thus recorded a positive trade balance of US $23 billion in 2006.

The main destinations for Chilean exports were the Americas (US $39 billion), Asia (US $27.8 billion) and Europe (US $22.2 billion). Seen as shares of Chile's export markets, 42% of exports went to the Americas, 30% to Asia and 24% to Europe. Within Chile's diversified network of trade relationships, its most important partner remained the United States. Total trade with the U.S. was US $14.8 billion in 2006. Since the U.S.-Chile Free Trade Agreement went into effect on January 1, 2004, U.S.-Chilean trade has increased by 154%. Internal Government of Chile figures show that even when factoring out inflation and the recent high price of copper, bilateral trade between the U.S. and Chile has grown over 60% since then.

Total trade with Europe also grew in 2006, expanding by 42%. The Netherlands and Italy were Chile's main European trading partners. Total trade with Asia also grew significantly at nearly 31%. Trade with Korea and Japan grew significantly, but China remained Chile's most important trading partner in Asia. Chile's total trade with China reached U.S. $8.8 billion in 2006, representing nearly 66% of the value of its trade relationship with Asia.

The growth of exports in 2006 was mainly caused by a strong increase in sales to the United States, the Netherlands, and Japan. These three markets alone accounted for an additional US $5.5 billion worth of Chilean exports. Chilean exports to the United States totaled US $9.3 billion, representing a 37.7% increase compared to 2005 (US $6.7 billion). Exports to the European Union were US $15.4 billion, a 63.7% increased compared to 2005 (US $9.4 billion). Exports to Asia increased from US $15.2 billion in 2005 to US $19.7 billion in 2006, a 29.9% increase.

During 2006, Chile imported US $26 billion from the Americas, representing 54% of total imports, followed by Asia at 22%, and Europe at 16%. Mercosur members were the main suppliers of imports to Chile at US $9.1 billion, followed by the United States with US $5.5 billion and the European Union with US $5.2 billion. From Asia, China was the most important exporter to Chile, with goods valued at US $3.6 billion. Year-on-year growth in imports was especially strong from a number of countries-Ecuador (123.9%), Thailand (72.1%), Korea (52.6%), and China (36.9%).

Chile's overall trade profile has traditionally been dependent upon copper exports. The state-owned firm CODELCO is the world's largest copper-producing company, with recorded copper reserves of 200 years. Chile has made an effort to expand nontraditional exports. The most important non-mineral exports are forestry and wood products, fresh fruit and processed food, fishmeal and seafood, and wine.

Successive Chilean governments have actively pursued trade-liberalizing agreements. During the 1990s, Chile signed free trade agreements (FTA) with Canada, Mexico, and Central America. Chile also concluded preferential trade agreements with Venezuela, Colombia, and Ecuador. An association agreement with Mercosur-Argentina, Brazil, Paraguay, and Uruguay-went into effect in October 1996. Continuing its export-oriented development strategy, Chile completed landmark free trade agreements in 2002 with the European Union and South Korea. Chile, as a member of the Asia-Pacific Economic Cooperation (APEC) organization, is seeking to boost commercial ties to Asian markets.

To that end, it has signed trade agreements in recent years with New Zealand, Singapore, Brunei, India, China, and most recently Japan. In 2007, Chile held trade negotiations with Australia, Thailand, Malaysia, and China. In 2008, Chile hopes to conclude an FTA with Australia, and finalize an expanded agreement (covering trade in services and investment) with China. The P4 (Chile, Singapore, New Zealand, and Brunei) also plan to expand ties through adding a finance and investment chapter to the existing P4 agreement. Chile's trade talks with Malaysia and Thailand are also scheduled to continue in 2008.

After two years of negotiations, the United States and Chile signed an agreement in June 2003 that will lead to completely duty-free bilateral trade within 12 years. The U.S.-Chile FTA entered into force January 1, 2004, following approval by the U.S. and Chilean congresses. The bilateral FTA has inaugurated greatly expanded U.S.-Chilean trade ties, with total bilateral trade jumping by 154% during the FTA's first three years.

Chile unilaterally lowered its across-the-board import tariff for all countries with which it does not have a trade agreement to 6% in 2003. Higher effective tariffs are charged only on imports of wheat, wheat flour, and sugar as a result of a system of import price bands. The price bands were ruled inconsistent with Chile's World Trade Organization (WTO) obligations in 2002, and the government has introduced legislation to modify them. Under the terms of the U.S.-Chile FTA, the price bands will be completely phased out for U.S. imports of wheat, wheat flour, and sugar within 12 years.

Chile is a strong proponent of pressing ahead on negotiations for a Free Trade Area of the Americas (FTAA) and is active in the WTO's Doha round of negotiations, principally through its membership in the G-20 and Cairns Group.

Chile is the first OECD member in South America.

Finance

Chile's financial sector has grown quickly in recent years, with a banking reform law approved in 1997 that broadened the scope of permissible foreign activity for Chilean banks. The Chilean Government implemented a further liberalization of capital markets in 2001, and there is further pending legislation proposing further liberalization. Over the last ten years, Chileans have enjoyed the introduction of new financial tools such as home equity loans, currency futures and options, factoring, leasing, and debit cards. The introduction of these new products has also been accompanied by an increased use of traditional instruments such as loans and credit cards. Chile's private pension system, with assets worth roughly $70 billion at the end of 2006, has been an important source of investment capital for the capital market. However, by 2009, it has been reported that $21 billion had been lost from the pension system to the global financial crisis. Chile maintains one of the best credit ratings (S&P A+) in Latin America. There are three main ways for Chilean firms to raise funds abroad: bank loans, issuance of bonds, and the selling of stocks on U.S. markets through American Depository Receipts (ADRs). Nearly all of the funds raised through these means go to finance domestic Chilean investment. The government is required by law to run a fiscal surplus of at least 1% of GDP. In 2006, the Government of Chile ran a surplus of $11.3 billion, equal to almost 8% of GDP. The Government of Chile continues to pay down its foreign debt, with public debt only 3.9% of GDP at the end of 2006.

Demographics

Chile's 2002 census reported a population of 15,116,435. Its rate of population growth has been decreasing since 1990, because of a declining birth rate. By 2050 the population is expected to reach approximately 20.2 million. About 85% of the country's population lives in urban areas, with 40% living in Greater Santiago. The largest agglomeration according to the 2002 census are Greater Santiago with 5.6 million people, Greater Concepcion with 861,000 and Greater Valparaiso with 824,000.

Ethnic Structure

Studies on the ethnic structure of Chile are inconclusive and might vary significantly from one study to the next.

One study conducted by Francisco Lizcano from UNAM considered that people of European origin made up 52.7% the population; Mestizos made up 44% of the population. A study conducted by the University of Chile found that within the Chilean population, 30% are of European descent; the Mestizo component of partly European ancestry is estimated to be 65% of the population. Other studies have found a white majority that would exceed 60% of the Chilean population.

The European segment of Chile's population consists mainly of people descended from Spanish settlers (predominantly Castilian and Basque), with important minorities of people of German, Italian, Irish, French, British, Swiss, and Croatian ancestry, singly or combined with each other. The Mestizo segment, in this respect, would derive its European component from colonial Spanish settlers (mainly Andalusians and Castilians), while its Amerindian component from various tribes or groups, mainly Picunches and Mapuches.

The Afro-Chilean population has always been tiny, reaching a high of 2,500 during the colonial period; their current percentage of the population is less than one percent. According to the Census 2002, 4.6% of the Chilean population considered themselves indigenous.

Indigenous Communities

The 1907 census reported 101,118 Indians, or 3.1% of the total country population. Only those that practiced their native culture or spoke their native language were considered, irrespective of their "racial purity."

At the 2002 census, only indigenous people that still practiced a native culture or spoke a native language were surveyed: 4.6% of the population (692,192 people) fit that description; of these, 87.3% declared themselves Mapuche. Most of the indigenous population show varying degrees of mixed ancestry.

Chile is one of the twenty countries to have signed and ratified the only binding international law concerning indigenous peoples, Indigenous and Tribal Peoples Convention, 1989. It was adopted in 1989 as the International Labour Organization Convention 169. Chile ratified the convention in 2008. In November 2009, a court decision in Chile, considered to be a landmark in indigenous rights concerns, made use of the ILO convention 169. The Supreme Court decision on Aymara water rights upholds rulings by both the Pozo Almonte tribunal and the Iquique Court of Appeals, and marks the first judicial application of ILO Convention 169 in Chile.

Immigration

A small, albeit influential, number of non-Spanish immigrants have arrived in Chile from various countries and regions, including Italy, Ireland, France, Greece, Germany, England, the Netherlands, Scotland, Croatia, and the Middle East. The largest contingent of people to arrive in Chile came from Spain and the Basque country, beginning in the sixteenth century. Estimates of the number of people in Chile who can trace descent from Basques range from 10% (1,600,000) to as high as 27% (4,500,000).

In 1848 an important and substantial German immigration took place, laying the foundation for the German-Chilean community. Sponsored by the Chilean government for the colonization of the southern region, the Germans (including German-speaking Swiss, Silesians, Alsatians and Austrians), strongly influenced the ethnic composition of the southern provinces of Chile. German immigrants have made a cultural impact in many areas of southern Chile, a sparsely populated region. The Consulate of Chile in Germany estimates that 500,000 to 600,000 Chileans, or between 3% to 3.5% of the population today, are descended from German immigrants.

It is estimated that nearly five percent of the Chilean population, or about 800,000 persons, are of Asian origin, chiefly from the Middle East (these include, most notably, Palestinians, Syrians, Lebanese and Middle Eastern Armenians). Note that Israelis, both Jewish and non-Jewish citizens of the nation of Israel, may be included. Chile is home to a large population of immigrants, mostly Christian, from the Levant. Roughly 500,000 of Chile's population is of full or partial Palestinian origin.

Other historically significant immigrant groups include Croatians, whose descendants today have been estimated to number some 380,000 persons (or about 2.4% of the population); other authorities estimate that close to 4.6% of the Chilean population may have some Croatian ancestry. Over 700,000 Chileans, or about 4.5% of the national population, may have some English, Scottish, or Welsh ancestry, being the highest porcentage in whole Latin America.

Chileans of Greek descent are estimated to number between 90,000 and 120,000 persons, placing Chile among the five countries in the world with the most descendants of Greeks.; most live in or near either Santiago or Antofagasta. Swiss descendants add another 90,000 to the population. Perhaps five percent of the Chilean population has some French ancestry. Between 600,000 and 800,000 Chileans are descended from Italian immigrants. Other groups of Europeans have followed but are found in smaller numbers, as the descendants of

Austrians and Dutchmen it is currently estimated at about 50,000. Altogether, these immigrants with their descendants, they have transformed the country culturally, economically and politically.

European emigration to Chile (and to a lesser extent, the arrivals from Middle East), during the second half of the nineteenth century and throughout the twentieth, was the most important in Latin America after that to the Atlantic Coast of the Southern Cone.

Descendants of different European ethnic groups often intermarried in Chile, diluting the cultures and separate identities of the home countries and fusing them together and with the descendants of the original *Basque-Castilian* aristocracy of the colonial period while at the same time preserving some separate aspects. This intermarriage and mixture has help shape the present society and culture of the Chilean middle and upper classes., who now enjoy varied elements of their original European cultures, such as British afternoon tea, German cakes, and Italian pasta. The fusion is visible in the architecture of Chilean cities. These classes do, however, frequently deprecate Chilean folk culture, an offshoot of the culture of the Spaniards who settled the country in the colonial period.

Chile has recently become a new magnet for immigrants, mostly from neighboring Argentina, Peru, and Bolivia. According to the 2002 national census, Chile's foreign-born foreign population has increased by 75% since 1992. According to an estimation by the Migration and Foreign Residency Department, 317,057 foreigners were living in Chile as of December 2008.

Religion

In the most recent census (2002), 70 percent of the population over age 14 identified as Roman Catholic and 15.1 percent as evangelical. In the census, the term "evangelical" referred to all non-Catholic Christian churches with the exception of the Orthodox Church (Greek, Persian, Serbian, Ukrainian, and Armenian), the Church of Jesus Christ of Latter-day Saints (Mormons), Seventh-day Adventists, and Jehovah's Witnesses. Approximately 90 percent of evangelicals are Pentecostal. Wesleyan, Lutheran, Reformed Evangelical,

Presbyterian, Anglican, Episcopalian, Baptist, and Methodist churches are also present. Irreligious people, atheists and agnostics, account for around 8% of the population.

The Constitution provides for freedom of religion, and other laws and policies contribute to the generally free practice of religion. The law at all levels protects this right in full against abuse, either by governmental or private actors.

Church and state are officially separate. The 1999 law on religion prohibits religious discrimination; however, the Catholic Church enjoys a privileged status and occasionally receives preferential treatment. Government officials attend Catholic events and also major Protestant and Jewish ceremonies.

The Government observes Christmas, Good Friday, the Feast of the Virgin of Carmen, the Feast of Saints Peter and Paul, the Feast of the Assumption, All Saints' Day, and the Feast of the Immaculate Conception as national holidays. The government has recently declared October 31, Reformation Day, a public national holiday, in honour of the Protestant churches of the country.

Languages

Spanish

The Spanish spoken in Chile is distinctively accented and quite unlike that of neighbouring South American countries because final syllables and "*s*" sounds are dropped, and some consonants have a soft pronunciation. Accent varies only very slightly from north to south; more noticeable are the small differences in accent based on social class or whether one lives in the city or the country. That the Chilean population largely formed in a small section of the center of the country and then migrated in modest numbers to the north and south helps explain this relative lack of differentiation, which was maintained by the national reach of radio and now television which diffuses and homogenizes colloquial expressions.

German

German is spoken to a limited extent in southern Chile, either in small country side pockets or as a second language among the communities of larger cities.

English

Through initiatives such as the English Opens Doors program, the government made English mandatory for fifth-graders and above in public schools while most private schools start teaching English from kindergarten. Common English words have been absorbed and appropriated into everyday Spanish speech.

Indigenous Languages

There are several indigenous languages spoken in Chile: Mapudungun, Quechua, Aymara and Rapa Nui. After the Spanish invasion, Spanish took over as the lingua franca and the indigenous languages have become minority languages, with some now extinct or close to extinction.

Government and Politics

The current Constitution of Chile was approved in a highly irregular national plebiscite in September 1980, under the military government of Augusto Pinochet. It entered into force in March 1981. After Pinochet's defeat in the 1988 plebiscite, the constitution was amended to ease provisions for future amendments to the Constitution. In September 2005, President Ricardo Lagos signed into law several constitutional amendments passed by Congress. These include eliminating the positions of appointed senators and senators for life, granting the President authority to remove the commanders-in-chief of the armed forces, and reducing the presidential term from six to four years. Chileans voted in the first round of presidential elections on December 13, 2009. None of the four presidential candidates got more than 50% of the vote. As a result, the top two candidates, center-left Cancerization de Partidos por la Democracia coalition's Eduardo Frei Ruiz-Tagle and center-right Coalition por el Cambio coalition's Sebastian Pinera, competed in a run-off election on January 17, 2010, which Sebastian Pinera won. This was Chile's fifth presidential election since the end of the Pinochet era. All five have been judged free and fair. The president is constitutionally barred from serving consecutive terms.

The Congress of Chile has a 38-seat Senate and a 120-member Chamber of Deputies. Senators serve for 8 years with staggered terms, while deputies are elected every 4 years. The current Senate has a 20-18 split in favour of the opposition coalition. The last congressional elections were held on December 13, 2009, concurrently with the presidential election. The current lower house-the Chamber of Deputies-contains 58 members of the governing center-right coalition, 54 from the center-left opposition and 8 from small parties or independents. The Congress is located in the port city of Valparaiso, about 140 kilometres (84 mi) west of the capital, Santiago.

Chile's congressional elections are governed by a binomial system that rewards the two largest representations. Therefore, there are only two senate and two deputy seats apportioned to each electoral district, parties are forced to form wide coalitions and, historically, the two largest coalitions (Cancerization and Alianza) split most of the seats in a district. Only if the leading coalition ticket out-polls the second place coalition by a margin of more than 2-to-1 does the winning coalition gain both seats.

In the 2001 congressional elections, the conservative Independent Democratic Union surpassed the Christian Democrats for the first time to become the largest party in the lower house. In the 2005 parliamentary election, both leading parties, the Christian Democrats and the UDI lost representation in favour of their respective allies Socialist Party (which became the biggest party in the Cancerization block) and National Renewal in the right-wing alliance. In the last legislative elections in Chile, the Communist Party won 3 out of 120 seats in the Chamber of Deputies for the first time in 30 years.

Chile's judiciary is independent and includes a court of appeal, a system of military courts, a constitutional tribunal, and the Supreme Court of Chile. In June 2005, Chile completed a nation-wide overhaul of its criminal justice system. The reform has replaced inquisitorial proceedings with an adversarial system more similar to that of the United States.

Chile is divided into 15 regions, each headed by an intendant appointed by the president. The regions are further divided into provinces, with provincial governors also appointed by the president. Finally each province is divided into communes which are administered by municipalities, each with its own mayor and council elected for four year terms. Each region is designated by a name and a Roman numeral, assigned from north to south. The only exception is the Santiago Metropolitan Region which is designated *RM* (*Region Metropolitana*). Two new regions were created in 2006 and became operative in October 2007; Los Rios in the south (Region XIV), and Arica y Parinacota in the north (Region XV). The numbering scheme skipped Region XIII.

Defense

Chile's Armed Forces are subject to civilian control exercised by the president through the Minister of Defense. The president has the authority to remove the commanders-in-chief of the armed forces.

The commander in chief of the Chilean Army is General Juan Miguel Fuente-Alba Poblete. The Chilean Army is 45,000 strong and is organized with an Army headquarters in Santiago, seven divisions throughout its territory, an Air Brigade in Rancagua, and a Special Forces Command in Colina. The Chilean Army is one of the most professional and technologically advanced armies in Latin America.

Admiral Edmundo Gonzalez Robles directs the 21,773-person Chilean Navy, including 2,500 Marines. Of the fleet of 29 surface vessels, only eight are operational major combatants (frigates). Those ships are based in Valparaiso. The Navy operates its own aircraft for transport and patrol; there are no Navy fighter or bomber aircraft. The Navy also operates four submarines based in Talcahuano.

Gen. Ricardo Ortega Perrier heads 12,500 strong Chilean Air Force. Air assets are distributed among five air brigades headquartered in Iquique, Antofagasta, Santiago, Puerto Montt, and Punta Arenas. The Air Force also operates an airbase on King George Island, Antarctica. The Air Force took delivery of the final 2 of 10 F-16s, all purchased from the U.S., in March 2007. Chile also took delivery in 2007 of a number of reconditioned Block 15 F-16s from the Netherlands, bringing to 18 the total of F-16s purchased from the Dutch.

After the military coup in September 1973, the Chilean national police (Carabineros) were incorporated into the Defense Ministry. With the return of democratic government, the police were placed under the operational control of the Interior Ministry but remained under the nominal control of the Defense Ministry. Gen. Eduardo Gordon is the head of the national police force of 40,964 men and women who are responsible for law enforcement, traffic management, narcotics suppression, border control, and counter-terrorism throughout Chile.

Foreign Relations

Since the early decades after independence, Chile has always had an active involvement in foreign affairs. In 1837 the country aggressively challenged the dominance of Peru's port of Callao for pre-eminence in the Pacific trade routes, defeating the short-lived alliance between Peru and Bolivia, the Peru-Bolivian Confederation (1836–39) in the War of the Confederation. The war dissolved the confederation while distributing power in the Pacific. A second international war, the War of the Pacific (1879–83), further increased Chile's regional role, while adding considerably to its territory.

During the nineteenth century, Chile's commercial ties were primarily with Britain, a country that had a decisive influence on the organization of the navy. The French influenced Chile's legal and educational systems and had a decisive impact on Chile, through the architecture of the capital in the boom years at the turn of the century. German influence came from the organization and training of the army by Prussians.

On June 26, 1945, Chile participated as a founding member of the United Nations being among 50 countries that signed the United Nations Charter in San Francisco, California. With the military coup of 1973, Chile became isolated politically as a result of widespread human rights abuses.

Since its return to democracy in 1990, Chile has been an active participant in the international political arena. Chile completed a 2-year non-permanent position on the UN Security Council in January 2005. Jose Miguel Insulza, a Chilean national, was elected Secretary General of the Organization of American States in May 2005. Chile is currently serving on the International Atomic Energy Agency (IAEA) Board of Governors, and the 2007-2008 chair of the board is Chile's ambassador to the IAEA, Milenko E. Skoknic. The country is an active member of the UN family of agencies and participates in UN peacekeeping activities. It is currently bidding for a seat on the UN Human Rights Council. Chile hosted the Defense Ministerial of the Americas in 2002 and the APEC summit and related meetings in 2004. It also hosted the Community of Democracies ministerial in April 2005 and the Ibero-American Summit in November 2007. An associate member of Mercosur and a full member of APEC, Chile has been an important actor on international economic issues and hemispheric free trade.

The Chilean Government has diplomatic relations with most countries. It settled its territorial disputes with Argentina during the 1990s. Chile and Bolivia severed diplomatic ties in 1978 over Bolivia's desire to reacquire territory it lost to Chile in 1879-83 War of the Pacific. The two countries maintain consular relations and are represented at the Consul General level.

Healthcare

The Ministry of Health of Chile (Spanish: Ministerio de Salud de Chile), also known as MINSAL, is the cabinet-level administrative office in charge of planning, directing, coordinating, executing, controlling and informing the public health policies formulated by the President of Chile.

Fondo Nacional de Salud also known as FONASA is the funding branch of the Ministry of Health (Chile). The National Health Fund (FONASA), created in 1979 by Decree Law No. 2763, is the financial entity entrusted to collect, manage and distribute state funds for health in Chile. It is funded by the public. Notably, all employees pay 7% of their monthly income to FONASA the funding branch of the Chilean Ministry of Health.

FONASA is part of the NHSS and has executive power through the Ministry of Health (Chile). FONASA headquarters are in Santiago and decentralized public service is conducted by various Regional Offices.

More than 12 million beneficiaries benefit from FONASA. Despite universal health care, beneficiaries can opt for more costly private insurance through ISAPRE. Political questions have been raised due to very wealthy beneficiaries opting for ISAPRE and thus generating healthcare inequity. Such is an example of socioeconomic disparity resulting in the high Gini coefficient in Chile.

Culture

During the period between early agricultural settlements and to the late pre-Hispanic period, northern Chile was a region of Andean culture that was influenced by altiplano traditions spreading to the coastal valleys of the north. While southern regions were areas of Mapuche cultural activities. Through the colonial period following the conquest, and during the early Republican period, the country's culture was dominated by the Spanish. Other European influences, primarily English, French, and German began in the 19th century and have continued to this day. German migrants influenced the Bavarian style rural architecture and cuisine in the south of Chile in cities such as Valdivia, Frutillar, Puerto Varas, Osorno, Temuco, Puerto Octay, Llanquihue, Faja Maisan, Pitrufquen, Victoria, Pucon and Puerto Montt.

Music and Dance

Music in Chile ranges from folkloric music, popular music and also to classical music. Its large geography generates different musical expressions in the north, center and south of the country, including also Easter Island and Mapuche music. The national dance is the cueca. Another form of traditional Chilean song, though not a dance, is the tonada. Arising from music imported by the Spanish colonists, it is distinguished from the cueca by an intermediate melodic section and a more prominent melody. Between 1950 and 1970 appears a rebirth in folk music leading by groups such as Los de Ramon, Los Cuatro Cuartos and Los Huasos Quincheros, among others with composers such as Raul de Ramon, Violeta Parra and others. In the mid-1960s native musical forms were revitalized by the Parra family with the Nueva Cancion Chilena, which was associated with political activists and reformers such as Victor Jara, in this period also appeared other important groups like Inti-Illimani. Other important folk singer and researcher on folklore and Chilean ethnography, is Margot Loyola. Also many Chilean Rock bands like La Ley, Los Tres, Los Prisioneros and Kudai have reached international success.

Literature

Chileans call their country *pais de poetas*-country of poets. Gabriela Mistral was the first Chilean to win a Nobel Prize for Literature (1945). Chile's most famous poet, however, is Pablo Neruda, who also won the Nobel Prize for Literature (1971) and is world-renowned for his extensive library of works on romance, nature, and politics. His three highly personalized homes, located in Isla Negra, Santiago and Valparaiso are popular tourist destinations.

Among the list of other Chilean poets are Carlos Pezoa Veliz, Vicente Huidobro, Gonzalo Rojas, Isabel Allende, and Nicanor Parra. Novelist Jose Donoso's novel *The Obscene Bird of the Night* is considered by critic Harold Bloom to be one of the canonical works of Twentieth Century Western literature. Another internationally recognized Chilean novelist is Roberto Bolano whose translations into English have had an excellent reception from the critics.

Cuisine

Chilean cuisine is a reflection of the country's topographical variety, featuring an assortment of seafood, beef, fruits, and vegetables. Traditional recipes include asado, cazuela, empanadas, humitas, pastel de choclo, pastel de papas and curanto.

Sports

Chile's most popular sport is association football. Chile has appeared in seven FIFA World Cups which includes hosting the 1962 FIFA World Cup where the national football team finished third. Other results achieved by the national football team include four finals at the Copa America, one silver and two bronze medals at the Pan American Games, a bronze medal at the 2000 Summer Olympics and two third places finishes in the FIFA under-17 and under-20 youth tournaments. The main football clubs are Colo-Colo, Universidad de Chile and Universidad Catolica. Colo-Colo is the country's most successful football club, having both the most national and international championships, including the coveted Copa Libertadores South American club tournament. Universidad Catolica was the last international champion (Interamerican

Cup 1994). Tennis is the country's most successful sport. Its national team won the World Team Cup clay tournament twice (2003 & 2004), and played the Davis Cup final against Italy in 1976. At the 2004 Summer Olympics the country captured gold and bronze in men's singles and gold in men's doubles. Marcelo Rios became the first Latin American man to reach the number one spot in the ATP singles rankings in 1998. Anita Lizana won the US Open in 1937, becoming the first woman from Latin America to win a grand slam tournament. Luis Ayala was twice a runner-up at the French Open and both Rios and Fernando Gonzalez reached the Australian Open men's singles finals.

At the Olympic Games Chile boasts two gold medals (tennis), seven silver medals (athletics, Equestrian, boxing, shooting and tennis) and four bronze medals (tennis, boxing and football). Rodeo is the country's national sport and is practiced in the more rural areas of the country. A sport similar to hockey called *chueca* was played by the Mapuche people during the Spanish conquest. Skiing and snowboarding are practiced at ski centres located in the Central Andes, and in southern ski centres near to cities as Osorno, Puerto Varas, Temuco and Punta Arenas. While surfing is popular at some coastal towns. Polo is professionally practiced within Chile and in 2008 Chile achieved top prize in the World Polo Championship a tournament where the country has earned both second and third places medals in previous editions. Basketball is a popular sport in which Chile has earned a bronze medal in the first men's FIBA World Championship held in 1950 and winning a second bronze medal when Chile hosted the 1959 FIBA World Championship. Chile hosted the first FIBA World Championship for Women in 1953 finishing the tournament with the silver medal. Other sports such as marathons and ultramarathons are also increasing in popularity. Chile's San Pedro de Atacama is host to the annual "Atacama Crossing," a six-stage, 250-kilometer footrace which annually attracts about 150 competitors from 35 countries.

Tourism

Tourism in Chile has experienced sustained growth over the last few decades. In 2005, tourism grew by 13.6%, generating more than 4.5 billion dollars of which 1.5 billion is attributed to foreign tourists. According to the National Service of Tourism (Sernatur), 2 million people a year visit the country. Most of these visitors come from other countries in the American continent, mainly Argentina; followed by a growing number from the United States, Europe, and Brazil with a growing number of Asians from South Korea and PR China.

The main attractions for tourists are places of natural beauty situated in the extreme zones of the country: San Pedro de Atacama, in the north, is very popular with foreign tourists who arrive to admire the Incaic architecture, the altiplano lakes, and the Valley of the Moon. In Putre, also in the North, there is the Chungara Lake, as well as the Parinacota and the Pomerape volcanoes, with altitudes of 6,348 m and 6,282 m, respectively.

Throughout the central Andes there are many ski resorts of international repute, like Portillo, Valle Nevado and Termas de Chillan . In the south, the main tourist sites are the coastal area around Tirua and Canete with the Isla Mocha and the Nahuelbuta National Park, Chiloe Archipelago and Patagonia, which includes Laguna San Rafael National Park, with its many glaciers, and the Torres del Paine National Park. The central port city of Valparaiso, with its unique architecture, is also popular. Finally, Easter Island in the Pacific Ocean is one of the main Chilean tourist destinations.

For locals, tourism is concentrated mostly in the summer (December to March), and mainly in the coastal beach towns. Arica, Iquique, Antofagasta, La Serena and Coquimbo are the main summer centres in the north, and Pucon on the shores of Lake Villarrica is the main one in the south. Because of its proximity to Santiago, the coast of the Valparaiso Region, with its many beach resorts, receives the largest number of tourists. Vina del Mar, Valparaiso's northern affluent neighbour, is popular because of its beaches, casino, and its annual song festival, the most important musical event in Latin America.

In November 2005, the government launched a campaign under the brand "Chile: All Ways Surprising," intended to promote the country internationally for both business and tourism.

National Symbols

The national flower is the copihue (*Lapageria rosea*, Chilean bellflower), which grows in the woods of southern Chile.

The coat of arms depicts the two national animals: the condor (*Vultur gryphus*, a very large bird that lives in the mountains) and the huemul (*Hippocamelus bisulcus*, an endangered white tail deer). It also has the legend *Por la razon o la fuerza* (*By reason or by force*).

The flag of Chile consists of two equal horizontal bands of white and red; there is a blue square the same height as the white band at the hoist-side end of the white band; the square bears a white five-pointed star in the center representing a guide to progress and honour; blue symbolizes the sky, white is for the snow-covered Andes, and red stands for the blood spilled to achieve independence.

9

Antarctica and Arctic

Antarctica is Earth's southernmost continent, underlying the South Pole. It is situated in the Antarctic region of the southern hemisphere, almost entirely south of the Antarctic Circle, and is surrounded by the Southern Ocean. At 14.0 million km (5.4 million sq mi), it is the fifth-largest continent in area after Asia, Africa, North America, and South America. About 98% of Antarctica is covered by ice, which averages at least 1.6 kilometres (1.0 mi) in thickness.

Antarctica, on average, is the coldest, driest, and windiest continent, and has the highest average elevation of all the continents. Antarctica is considered a desert, with annual precipitation of only 200 mm (8 inches) along the coast and far less inland. There are no permanent human residents but anywhere from 1,000 to 5,000 people reside throughout the year at the research stations scattered across the continent. Only cold-adapted plants and animals survive there, including penguins, seals, nematodes, Tardigrades, mites, many types of algae and other microorganisms, and tundra vegetation.

Although myths and speculation about a *Terra Australis* ("Southern Land") date back to antiquity, the first confirmed sighting of the continent is commonly accepted to have occurred in 1820 by the Russian expedition of Mikhail Lazarev and Fabian Gottlieb von Bellingshausen. The continent, however, remained largely neglected for the rest of the 19th century because of its hostile environment, lack of resources, and isolation. The first formal use of the name "Antarctica" as a continental name in the 1890s is attributed to the Scottish cartographer John George Bartholomew. The Antarctic Treaty was signed in 1959 by twelve countries; to date, forty-six countries have signed the treaty. The treaty prohibits military activities and mineral mining, supports scientific research, and protects the continent's ecozone. Ongoing experiments are conducted by more than 4,000 scientists of many nationalities and with various research interests.

History

Belief in the existence of a *Terra Australis* — a vast continent in the far south of the globe to "balance" the northern lands of Europe, Asia and North Africa — has existed since the times of Ptolemy (1st century AD), who suggested the idea to preserve the symmetry of all known landmasses in the world. Depictions of a large southern landmass were common in maps such as the early 16th century Turkish Piri Reis map. Even in the late 17th century, after explorers had found that South America and Australia were not part of the fabled "Antarctica", geographers believed that the continent was much larger than its actual size.

European maps continued to show this hypothetical land until Captain James Cook's ships, HMS *Resolution* and *Adventure*, crossed the Antarctic Circle on 17 January 1773, in December 1773 and again in January 1774. Cook in fact came within about 75 miles (121 km) of the Antarctic coast before retreating in the face of field ice in January 1773. The first confirmed sighting of Antarctica can be narrowed down to the crews of ships captained by three individuals. According to various organizations (the National Science Foundation, NASA, the University of California, San Diego, and other sources), ships captained by three men sighted Antarctica in 1820: Fabian Gottlieb von Bellingshausen (a captain in the Russian Imperial Navy), Edward Bransfield (a captain in the Royal Navy), and Nathaniel Palmer (an American sealer out of Stonington, Connecticut). Von Bellingshausen saw Antarctica on 27 January 1820, three days before Bransfield sighted land, and ten months before Palmer did so in November 1820. On that day the two-ship expedition led by Von Bellingshausen and Mikhail Petrovich Lazarev reached a point within 32 kilometers (20 mi) of the Antarctic mainland and saw ice fields there. The first documented landing on mainland Antarctica was by the American sealer John Davis in West Antarctica on 7 February 1821, although some historians dispute this claim.

In December, 1839, as part of the United States Exploring Expedition of 1838–42 conducted by the United States Navy (sometimes called the "Ex. Ex.", or "the Wilkes Expedition"), an expedition sailed from Sydney, Australia, into the Antarctic Ocean, as it was then known, and reported the discovery "of an Antarctic continent west of the Balleny Islands". That part of Antarctica was later named "Wilkes Land", a name it maintains to this day.

Explorer James Clark Ross passed through what is now known as the Ross Sea and discovered Ross Island (both of which were named for him) in 1841. He sailed along a huge wall of ice that was later named the Ross Ice Shelf (also named for him). Mount Erebus and Mount Terror are named after two ships from his expedition: HMS *Erebus* and *Terror*. Mercator Cooper landed in East Antarctica on 26 January 1853.

During the Nimrod Expedition led by Ernest Shackleton in 1907, parties led by T. W. Edgeworth David became the first to climb Mount Erebus and to reach the South Magnetic Pole. Douglas Mawson, who assumed the leadership of the Magnetic Pole party on their perilous return, went on to lead several expeditions until retiring in 1931. In addition, Shackleton himself and three other members of his expedition made several firsts in December 1908 – February 1909: they were the first humans to traverse the Ross Ice Shelf, the first to traverse the Transantarctic Mountain Range (via the Beardmore Glacier), and the first to set foot on the South Polar Plateau. An expedition led by Norwegian polar explorer Roald Amundsen from the ship *Fram* became the first to reach the geographic South Pole on 14 December 1911, using a route from the Bay of Whales and up the Axel Heiberg Glacier. One month later, the ill-fated Scott Expedition reached the pole.

Richard Evelyn Byrd led several voyages to the Antarctic by plane in the 1930s and 1940s. He is credited with implementing mechanized land transport on the continent and conducting extensive geological and biological research. However, it was not until 31 October 1956 that anyone set foot on the South Pole again; on that day a U.S. Navy group led by Rear Admiral George J. Dufek successfully landed an aircraft there.

The first person to sail single-handed to Antarctica was the New Zealander David Henry Lewis, in a 10-meter steel sloop *Ice Bird*.

Geography

Entered asymmetrically around the South Pole and largely south of the Antarctic Circle, Antarctica is the southernmost continent and is surrounded by the Southern Ocean; alternatively, it may be considered to be surrounded by the southern Pacific, Atlantic, and Indian Oceans, or by the southern waters of the World Ocean. It covers more than 14,000,000 km (5,400,000 sq mi), making it the fifth-largest continent, about 1.3 times as large as Europe. The coastline measures 17,968 km (11,165 mi) and is mostly characterized by ice formations, as the following table shows:

Coastal Types Around Antarctica (Drewry, 1983)

Type *Frequency*	
Ice shelf (floating ice front)	44%
Ice walls (resting on ground)	38%
Ice stream/outlet glacier (ice front or ice wall)	13%
Rock	5%
Total	100%

Antarctica is divided in two by the Transantarctic Mountains close to the neck between the Ross Sea and the Weddell Sea. The portion west of the Weddell Sea and east of the Ross Sea is called West Antarctica and the remainder East Antarctica, because they roughly correspond to the Western and Eastern Hemispheres relative to the Greenwich meridian.

About 98% of Antarctica is covered by the Antarctic ice sheet, a sheet of ice averaging at least 1.6 km (1.0 mi) thick. The continent has about 90% of the world's ice (and thereby about 70% of the world's fresh water). If all of this ice were melted, sea levels would rise about 60 m (200 ft). In most of the interior of the continent, precipitation is very low, down to 20 mm (0.8 in) per year; in a few "blue ice" areas precipitation is lower than mass loss by sublimation and so the local mass balance is negative. In the dry valleys the same effect occurs over a rock base, leading to a desiccated landscape.

West Antarctica is covered by the West Antarctic Ice Sheet. The sheet has been of recent concern because of the real, if small, possibility of its collapse. If the sheet were to break down, ocean levels would rise by several metres in a relatively geologically short period of time, perhaps a matter of centuries. Several Antarctic ice streams, which account for about 10% of the ice sheet, flow to one of the many Antarctic ice shelves.

East Antarctica lies on the Indian Ocean side of the Transantarctic Mountains and comprises Coats Land, Queen Maud Land, Enderby Land, Mac Robertson Land, Wilkes Land and Victoria Land. All but a small portion of this region lies within the Eastern Hemisphere. East Antarctica is largely covered by the East Antarctic Ice Sheet.

Vinson Massif, the highest peak in Antarctica at 4,892 m (16,050 ft), is located in the Ellsworth Mountains. Antarctica contains many other mountains, both on the main continent and the surrounding islands. Located on Ross Island, Mount Erebus is the world's southernmost active volcano. Another well-known volcano is found on Deception Island, which is famous for a giant eruption in 1970. Minor eruptions are frequent and lava flow has been observed in recent years. Other dormant volcanoes may potentially be active. In 2004, an underwater volcano was found in the Antarctic Peninsula by American and Canadian researchers. Recent evidence shows this unnamed volcano may be active.

Antarctica is home to more than 70 lakes that lie at the base of the continental ice sheet. Lake Vostok, discovered beneath Russia's Vostok Station in 1996, is the largest of these subglacial lakes. It was once believed that the lake had been sealed off for 500,000 to one million years but a recent survey suggests that, every so often, there are large flows of water from one lake to another.

There is some evidence, in the form of ice cores drilled to about 400 m (1,300 ft) above the water line, that Lake Vostok's waters may contain microbial life. The frozen surface of the lake shares similarities with Jupiter's moon Europa. If life is discovered in Lake Vostok, this would strengthen the argument for the possibility of life on Europa. On 7 February 2008, a NASA team embarked on a mission to Lake Untersee, searching for extremophiles in its highly alkaline waters. If found, these resilient creatures could further bolster the argument for extraterrestrial life in extremely cold, methane-rich environments.

Geology

Geological History and Paleontology

More than 170 million years ago, Antarctica was part of the supercontinent Gondwana. Over time, Gondwana gradually broke apart and Antarctica as we know it today was formed around 25 million years ago.

Paleozoic Era (540–250 Ma)

During the Cambrian period, Gondwana had a mild climate. West Antarctica was partially in the Northern Hemisphere, and during this period large amounts of sandstones, limestones and shales were deposited. East Antarctica was at the equator, where sea floor invertebrates and trilobites flourished in the tropical seas. By the start of the Devonian period (416 Ma), Gondwana was in more southern latitudes and the climate was cooler, though fossils of land plants are known from this time. Sand and silts were laid down in what is now the Ellsworth, Horlick and Pensacola Mountains. Glaciation began at the end of the Devonian period (360 Ma), as Gondwana became entered around the South Pole and the climate cooled, though flora remained. During the Permian period, the plant life became dominated by fern-like plants such as *Glossopteris*, which grew in swamps. Over time these swamps became deposits of coal in the Transantarctic Mountains. Towards the end of the Permian period, continued warming led to a dry, hot climate over much of Gondwana.

Mesozoic Era (250–65 Ma)

As a result of continued warming, the polar ice caps melted and much of Gondwana became a desert. In Eastern Antarctica, the seed fern became established, and large amounts of sandstone and shale were laid down at this time. Synapsids, commonly known as "mammal-like reptiles", were common in Antarctica during the Late Permian and Early Triassic and included forms such as *Lystrosaurus*. The Antarctic Peninsula began to form during the Jurassic period (206–146 Ma), and islands gradually rose out of the ocean. Ginkgo trees and cycads were plentiful during this period. In West Antarctica, coniferous forests dominated through the entire Cretaceous period (146–65 Ma), though Southern beech began to take over at the end of this period. Ammonites were common in the seas around Antarctica, and dinosaurs were also present, though only three Antarctic dinosaur genera (*Cryolophosaurus* and *Glacialisaurus*, from the Hanson Formation, and *Antarctopelta*) have been described to date. It was during this period that Gondwana began to break up.

Gondwanaland Breakup (160–23 Ma)

The cooling of Antarctica occurred stepwise, as the continental spread changed the oceanic currents from longitudinal equator-to-pole temperature-equalizing currents to latitudinal currents that preserved and accentuated latitude temperature differences.

Africa separated from Antarctica around 160 Ma, followed by the Indian subcontinent, in the early Cretaceous (about 125 Ma). About 65 Ma, Antarctica (then connected to Australia) still had a tropical to subtropical climate, complete with a marsupial fauna. About 40 Ma Australia-New Guinea separated from Antarctica, so that latitudinal currents could isolate Antarctica from Australia, and the first ice began to appear. During the Eocene-Oligocene extinction event about 34 million years ago, CO_2 levels have been found to be about 760 ppm and had been decreasing from earlier levels in the thousands of ppm.

Around 23 Ma, the Drake Passage opened between Antarctica and South America, resulting in the Antarctic Circumpolar Current that completely isolated the continent. Models of the changes suggest that declining CO_2 levels became more important. The ice began to spread, replacing the forests that then covered the continent. Since about 15 Ma, the continent has been mostly covered with ice, with the Antarctic ice cap reaching its present extension around 6 Ma.

Geology of Present-day Antarctica

The geological study of Antarctica has been greatly hindered by the fact that nearly all of the continent is permanently covered with a thick layer of ice. However, new techniques such as remote sensing, ground-penetrating radar and satellite imagery have begun to reveal the structures beneath the ice.

Geologically, West Antarctica closely resembles the Andes mountain range of South America. The Antarctic Peninsula was formed by uplift and metamorphism of sea bed sediments during the late Paleozoic and the early Mesozoic eras. This sediment uplift was accompanied by igneous intrusions and volcanism. The most common rocks in West Antarctica are andesite and rhyolite volcanics formed during the Jurassic period. There is also evidence of volcanic activity, even after the ice sheet had formed, in Marie Byrd Land and Alexander Island. The only anomalous area of West Antarctica is the Ellsworth Mountains region, where the stratigraphy is more similar to the eastern part of the continent.

East Antarctica is geologically varied, dating from the Precambrian era, with some rocks formed more than 3 billion years ago. It is composed of a metamorphic and igneous platform which is the basis of the continental shield. On top of this base are various modern rocks, such as sandstones, limestones, coal and shales laid down during the Devonian and Jurassic periods to form the Transantarctic Mountains. In coastal areas such as Shackleton Range and Victoria Land some faulting has occurred. The main mineral resource known on the continent is coal. It was first recorded near the Beardmore Glacier by Frank Wild on the Nimrod Expedition, and now low-grade coal is known across many parts of the Transantarctic Mountains. The Prince Charles Mountains contain significant deposits of iron ore. The most valuable resources of Antarctica lie offshore, namely the oil and natural gas fields found in the Ross Sea in 1973. Exploitation of all mineral resources is banned until 2048 by the Protocol on Environmental Protection to the Antarctic Treaty.

Climate

Antarctica is the coldest place on Earth. The coldest natural temperature ever recorded on Earth was 89.2 °C (128.6 °F) at the Russian Vostok Station in Antarctica on 21 July 1983. For comparison, this is 11 °C (20 °F) colder than subliming dry ice. Antarctica is a frozen desert with little precipitation; the South Pole itself receives less than 10 cm (4 in) per year, on average. Temperatures reach a minimum of between 80 °C (112 °F) and 90 °C ("130 °F) in the interior in winter and reach a maximum of between 5 °C (41 °F) and 15 °C (59 °F) near the coast in summer. Sunburn is often a health issue as the snow surface reflects almost all of the ultraviolet light falling on it.

East Antarctica is colder than its western counterpart because of its higher elevation. Weather fronts rarely penetrate far into the continent, leaving the center cold and dry. Despite the lack of precipitation over the central portion of the continent, ice there lasts for extended time periods. Heavy snowfalls are not uncommon on the coastal portion of the continent, where snowfalls of up to 1.22 metres (48 in) in 48 hours have been recorded.

At the edge of the continent, strong katabatic winds off the polar plateau often blow at storm force. In the interior, however, wind speeds are typically moderate. During summer, more solar radiation reaches the surface during clear days at the South Pole than at the equator because of the 24 hours of sunlight each day at the Pole.

Antarctica is colder than the Arctic for two reasons. First, much of the continent is more than 3 kilometres (2 mi) above sea level, and temperature decreases with elevation. Second, the Arctic Ocean covers the north polar zone: the ocean's relative warmth is transferred through the icepack and prevents temperatures in the Arctic regions from reaching the extremes typical of the land surface of Antarctica.

Given the latitude, long periods of constant darkness or constant sunlight create climates unfamiliar to human beings in much of the rest of the world. The aurora australis, commonly known as the southern lights, is a glow observed in the night sky near the South Pole created by the plasma-full solar winds that pass by the Earth. Another unique spectacle is diamond dust, a ground-level cloud composed of tiny ice crystals. It generally forms under otherwise clear or nearly clear skies, so people sometimes also refer to it as clear-sky precipitation. A sun dog, a frequent atmospheric optical phenomenon, is a bright "spot" beside the true sun.

Population

Antarctica has no permanent residents, but a number of governments maintain permanent manned research stations throughout the continent. The number of people conducting and supporting scientific research and other work on the continent and its nearby islands varies from about 1,000 in winter to about 5,000 in the summer. Many of the stations are staffed year-round, the winter-over personnel typically arriving from their home countries for a one-year assignment. An Orthodox church opened in 2004 at the Russian Bellingshausen Station is also manned year-round by one or two priests, who are similarly rotated every year.

The first semi-permanent inhabitants of regions near Antarctica (areas situated south of the Antarctic Convergence) were British and American sealers who used to spend a year or more on South Georgia, from 1786 onward. During the whaling era, which lasted until 1966, the population of that island varied from over 1,000 in the summer (over 2,000 in some years) to some 200 in the winter. Most of the whalers were Norwegian, with an increasing proportion of Britons. The settlements included Grytviken, Leith Harbour, King Edward Point, Stromness, Husvik, Prince Olav Harbour, Ocean Harbour and Godthul. Managers and other senior officers of the whaling stations often lived together with their families. Among them was the founder of Grytviken, Captain Carl Anton Larsen, a prominent Norwegian whaler and explorer who, along with his family, adopted British citizenship in 1910.

The first child born in the southern polar region was Norwegian girl Solveig Gunbjorg Jacobsen, born in Grytviken on 8 October 1913, and her birth was registered by the resident British Magistrate of South Georgia. She was a daughter of Fridthj of Jacobsen, the assistant manager of the whaling station, and of Klara Olette Jacobsen. Jacobsen arrived on the island in 1904 to become the manager of Grytviken, serving from 1914 to 1921; two of his children were born on the island. Emilio Marcos Palma was the first person born south of the 60th parallel south (the continental limit according to the Antarctic Treaty), as well as the first one born on the Antarctic mainland, in 1978 at Base Esperanza, on the tip of the Antarctic Peninsula; his parents were sent there along with seven other families by the Argentine government to determine if family life was suitable on the continent. In 1984, Juan Pablo Camacho was born at the Frei Montalva Station, becoming the first Chilean born in Antarctica. Several bases are now home to families with children attending schools at the station. As of 2009, eleven children were born in Antarctica (south of the 60th parallel south): eight at the Argentinean Esperanza Base and three at the Chilean Frei Montalva Station.

Flora and fauna

Flora

The climate of Antarctica does not allow extensive vegetation. A combination of freezing temperatures, poor soil quality, lack of moisture, and lack of sunlight inhibit plant growth. As a result, plant life is limited to mostly mosses and liverworts. The autotrophic community is made up of mostly protists. The flora of the continent largely consists of lichens, bryophytes, algae, and fungi. Growth generally occurs in the summer, and only for a few weeks at most.

There are more than 200 species of lichens and about 50 species of bryophytes, such as mosses. Seven hundred species of algae exist, most of which are phytoplankton. Multicolored snow algae and diatoms are especially abundant in the coastal regions during the summer. There are two species of flowering plants found in the Antarctic Peninsula: *Deschampsia antarctica* (Antarctic hair grass) and *Colobanthus quitensis* (Antarctic pearlwort).

Fauna

Few terrestrial vertebrates live in Antarctica. Invertebrate life includes microscopic mites like the Alaskozetes antarcticus, lice, nematodes, tardigrades, rotifers, krill and springtails. Recently ancient ecosystems consisting of several types of bacteria have been found living trapped deep beneath glaciers. The flightless midge *Belgica antarctica*, just 12 millimeters (0.5 in) in size, is the largest purely terrestrial animal in

Antarctica. The Snow Petrel is one of only three birds that breed exclusively in Antarctica.

A variety of marine animals exist and rely, directly or indirectly, on the phytoplankton. Antarctic sea life includes penguins, blue whales, orcas, colossal squids and fur seals. The Emperor penguin is the only penguin that breeds during the winter in Antarctica, while the Adelie Penguin breeds farther south than any other penguin. The Rockhopper penguin has distinctive feathers around the eyes, giving the appearance of elaborate eyelashes. King penguins, Chinstrap penguins, and Gentoo Penguins also breed in the Antarctic.

The Antarctic fur seal was very heavily hunted in the 18th and 19th centuries for its pelt by sealers from the United States and the United Kingdom. The Weddell Seal, a "true seal", is named after Sir James Weddell, commander of British sealing expeditions in the Weddell Sea. Antarctic krill, which congregates in large schools, is the keystone species of the ecosystem of the Southern Ocean, and is an important food organism for whales, seals, leopard seals, fur seals, squid, icefish, penguins, albatrosses and many other birds.

The passing of the Antarctic Conservation Act in the U.S. brought several restrictions to U.S. activity on Antarctica. The introduction of alien plants or animals can bring a criminal penalty, as can the extraction of any indigenous species. The over fishing of krill, which plays a large role in the Antarctic ecosystem, led officials to enact regulations on fishing. The Convention for the Conservation of Antarctic Marine Living Resources (CCAMLR), a treaty that came into force in 1980, requires that regulations managing all Southern Ocean fisheries consider potential effects on the entire Antarctic ecosystem. Despite these new acts, unregulated and illegal fishing, particularly of Patagonian toothfish (marketed as Chilean Sea Bass in the U.S.), remains a serious problem. The illegal fishing of toothfish has been increasing, with estimates of 32,000 tonnes (35,300 short tons) in 2000.

A census of sea life carried out during the International Polar Year and which involved some 500 researchers is due for release in 2010. The research is part of the global Census of Marine Life (CoML) and has disclosed some remarkable findings.

More than 235 marine organisms live in both polar regions, having bridged the gap of 12,000 km (7,456 mi). Large animals such as some cetaceans and birds make the round trip annually. More surprising are small forms of life such as mudworms, sea cucumbers and free-swimming snails found in both polar oceans. Various factors may aid in their distribution - fairly uniform temperatures of the deep ocean at the poles and the equator which differ by no more than 5 °C, and the major current systems or marine conveyor belt which transport egg and larvae stages.

Politics

Antarctica has no government, although various countries claim sovereignty in certain regions. While a few of these countries have mutually recognised each other's claims, the validity of these claims are generally not recognised universally.

New claims on Antarctica have been suspended since 1959 and the continent is considered politically neutral. Its status is regulated by the 1959 Antarctic Treaty and other related agreements, collectively called the Antarctic Treaty System. Antarctica is defined as all land and ice shelves south of 60° S for the purposes of the Treaty System. The treaty was signed by twelve countries including the Soviet Union (and later Russia), the United Kingdom, Argentina, Chile, Australia, and the United States. It set aside Antarctica as a scientific preserve, established freedom of scientific investigation and environmental protection, and banned military activity on the continent. This was the first arms control agreement established during the Cold War.

In 1983, the Antarctic Treaty Parties began negotiations on a convention to regulate mining in Antarctica. A coalition of international organisations launched a public pressure campaign to prevent any minerals development in the region, led largely by Greenpeace International which established its own scientific station–World Park Base–in the Ross Sea region and conducted annual expeditions to document environmental effects of humans on the continent. In 1988, the Convention on the Regulation of Antarctic Mineral Resources (CRAMRA) was adopted. The following year, however, Australia and France announced that they would not ratify the convention, rendering it dead for all intents and purposes. They proposed instead that a comprehensive regime to protect the Antarctic environment be negotiated in its place. The Protocol on Environmental Protection to the Antarctic Treaty (the 'Madrid Protocol') was negotiated as other countries followed suit and on 14 January 1998 it entered into force. The Madrid Protocol bans all mining in Antarctica, designating the continent as a 'natural reserve devoted to peace and science'.

The Antarctic Treaty prohibits any military activity in Antarctica, including the establishment of military bases and fortifications, military manoeuvers, and weapons testing. Military personnel or equipment are permitted only for scientific research or other peaceful purposes. The only documented military land manoeuvre was Operation NINETY by the Argentine military. The United States military issues the Antarctica Service Medal to military members or civilians who perform research duty in Antarctica. The medal includes a "wintered over" bar issued to those who remain on the continent for 2 six-month seasons.

Antarctic Territories

The Argentine, British, and Chilean claims all overlap, and have caused friction. The areas shown as Australia's and New Zealand's claims were British territory until they were handed over following the countries' independence. Australia currently claims the largest area. Australia and New Zealand both recognise the British claim, and vice-versa.

Countries Interested in Participating in a Possible Territorial Division of Antarctica

This group of countries participating as members of Antarctica Treaty have a territorial interest in the Antarctic continent but the provisions of the Treaty do not allow them to make their claims while it is in force.

- Brazil has a designated 'zone of interest' that is not an actual claim.
- Peru has formally reserved its right to make a claim.
- Russia has reserved its right to claim "territories discovered by Russians", which potentially may refer to the entire continent.

- South Africa has formally reserved its right to make a claim.
- Spain has formally reserved its right to make a claim.
- United States has formally reserved its right to make a claim.

Economy

Although coal, hydrocarbons, iron ore, platinum, copper, chromium, nickel, gold and other minerals have been found, they have not been in large enough quantities to exploit. The 1991 Protocol on Environmental Protection to the Antarctic Treaty also restricts a struggle for resources. In 1998, a compromise agreement was reached to place an indefinite ban on mining, to be reviewed in 2048, further limiting economic development and exploitation. The primary economic activity is the capture and offshore trading of fish. Antarctic fisheries in 2000–01 reported landing 112,934 tonnes.

Small-scale "expedition tourism" has existed since 1957 and is currently subject to Antarctic Treaty and Environmental Protocol provisions, but in effect self-regulated by the International Association of Antarctica Tour Operators (IAATO). Not all vessels associated with Antarctic tourism are members of IAATO, but IAATO members account for 95% of the tourist activity. Travel is largely by small or medium ship, focusing on specific scenic locations with accessible concentrations of iconic wildlife. A total of 37,506 tourists visited during the 2006–07 Austral summer with nearly all of them coming from commercial ships. The number is predicted to increase to over 80,000 by 2010.

There has been some recent concern over the potential adverse environmental and ecosystem effects caused by the influx of visitors. A call for stricter regulations for ships and a tourism quota has been made by some environmentalists and scientists. The primary response by Antarctic Treaty Parties has been to develop, through their Committee for Environmental Protection and in partnership with IAATO, "site use guidelines" setting landing limits and closed or restricted zones on the more frequently visited sites. Antarctic sight seeing flights (which did not land) operated out of Australia and New Zealand until the fatal crash of Air New Zealand Flight 901 in 1979 on

Mount Erebus, which killed all 257 aboard. Qantas resumed commercial overflights to Antarctica from Australia in the mid-1990s.

Transport

Transport on the continent has transformed from explorers crossing the isolated remote area of Antarctica on foot to a more open area due to human technologies enabling more convenient and faster transport by land and predominantly by air and water. The use of dogs to pull researchers and sledges has been banned on objections that dogs are an alien species to Antarctica and menaces to wildlife as superpredators.

Research

Each year, scientists from 27 different nations conduct experiments not reproducible in any other place in the world. In the summer more than 4,000 scientists operate research stations; this number decreases to nearly 1,000 in the winter. McMurdo Station is capable of housing more than 1,000 scientists, visitors, and tourists.

Researchers include biologists, geologists, oceanographers, physicists, astronomers, glaciologists, and meteorologists. Geologists tend to study plate tectonics, meteorites from outer space, and resources from the breakup of the supercontinent Gondwanaland. Glaciologists in Antarctica are concerned with the study of the history and dynamics of floating ice, seasonal snow, glaciers, and ice sheets. Biologists, in addition to examining the wildlife, are interested in how harsh temperatures and the presence of people affect adaptation and survival strategies in a wide variety of organisms. Medical physicians have made discoveries concerning the spreading of viruses and the body's response to extreme seasonal temperatures. Astrophysicists at Amundsen-Scott South Pole Station study the celestial dome and cosmic microwave background radiation. Many astronomical observations are better made from the interior of Antarctica than from most surface locations because of the high elevation, which results in a thin atmosphere, low temperature, which minimizes the amount of water vapour in the atmosphere, and absence of light pollution, thus allowing for a view of space clearer than anywhere else on Earth. Antarctic ice serves as both the shield and the detection medium for the largest neutrino telescope in the world, built 2 kilometers below Amundsen-Scott station.

Since the 1970s, an important focus of study has been the ozone layer in the atmosphere above Antarctica. In 1985, three British Scientists working on data they had gathered at Halley Station on the Brunt Ice Shelf discovered the existence of a hole in this layer. In 1998, NASA satellite data showed that the Antarctic ozone hole was the largest on record, covering 27 million km (10 million sq mi).

It was eventually determined that the destruction of the ozone was caused by chlorofluorocarbons emitted by human products. With the ban of CFCs in the Montreal Protocol of 1989, it is believed that the ozone hole will close up over the next fifty years.

Princess Elisabeth Polar Science Station

On 6 September 2007, Belgian-based International Polar Foundation unveiled the Princess Elisabeth station, the world's first zero-emissions polar science station in Antarctica to research climate change. Costing $16.3 million, the prefabricated station, which is part of International Polar Year, was shipped to the South Pole from Belgium by the end of 2008 to monitor the health of the polar regions. Belgian polar explorer Alain Hubert stated: "This base will be the first of its kind to produce zero emissions, making it a unique model of how energy should be used in the Antarctic." Johan Berte is the leader of the station design team and manager of the project which conducts research in climatology, glaciology and microbiology.

Meteorites

Meteorites from Antarctica are an important area of study of material formed early in the solar system; most are thought to come from asteroids, but some may have originated on larger planets. The first meteorites were found in 1912. In 1969, a Japanese expedition discovered nine meteorites. Most of these meteorites have fallen onto the ice sheet in the last million years. Motion of the ice sheet tends to concentrate the meteorites at blocking locations such as mountain ranges, with wind erosion bringing them to the surface after centuries beneath accumulated snowfall. Compared with meteorites collected in more temperate regions on Earth, the Antarctic meteorites are well-preserved.

This large collection of meteorites allows a better understanding of the abundance of meteorite types in the solar system and how meteorites relate to asteroids and comets. New types of meteorites and rare meteorites have been found. Among these are pieces blasted off the Moon, and probably Mars, by impacts. These specimens, particularly ALH84001 discovered by ANSMET, are at the center of the controversy about possible evidence of microbial life on Mars. Because meteorites in space absorb and record cosmic radiation, the time elapsed since the meteorite hit the Earth can be determined from laboratory studies. The elapsed time since fall, or terrestrial residence age, of a meteorite represents more information that might be useful in environmental studies of Antarctic ice sheets.

In 2006, a team of researchers from Ohio State University used gravity measurements by NASA's GRACE satellites to discover the 300-mile (480 km)-wide Wilkes Land crater, which probably formed about 250 million years ago.

Volcanic Eruption

In January 2008, the British Antarctic Survey (BAS) scientists, led by Hugh Corr and David Vaughan, reported (in the journal Nature Geoscience) that 2,200 years ago, a volcano erupted under Antarctica ice sheet (based on airborne survey with radar images). The biggest eruption in Antarctica in the last 10,000 years, the volcanic ash was found deposited on the ice surface under the Hudson Mountains, close to Pine Island Glacier.

Ice Mass and Global Sea Level

Due to its location at the South Pole, Antarctica receives relatively little solar radiation. This means that it is a very cold continent where water is mostly in the form of ice. Precipitation is low (most of Antarctica is a desert) and almost always in the form of snow, which accumulates and forms a giant ice sheet which covers the land. Parts of this ice sheet form moving glaciers known as ice streams, which flow towards the edges of the continent. Next to the continental shore are many ice shelves. These are floating extensions of outflowing glaciers from the continental ice mass. Offshore, temperatures are also low enough that ice is formed from seawater through most of the year. It is important to understand the various types of Antarctic ice to understand possible effects on sea levels and the implications of global warming.

Sea ice extent expands annually in the Antarctic winter and most of this ice melts in the summer. This ice is formed from the ocean water and floats in the same water and thus does not contribute to rise in sea level. The extent of sea ice around Antarctica has remained roughly constant in recent decades, although the thickness changes are unclear.

Melting of floating ice shelves (ice that originated on the land) does not in itself contribute much to sea-level rise (since the ice displaces only its own mass of water). However it is the outflow of the ice from the land to form the ice shelf which causes a rise in global sea level. This effect is offset by snow falling back onto the continent. Recent decades have witnessed several dramatic collapses of large ice shelves around the coast of Antarctica, especially along the Antarctic Peninsula. Concerns have been raised that disruption of ice shelves may result in increased glacial outflow from the continental ice mass.

On the continent itself, the large volume of ice present stores around 70% of the world's fresh water. This ice sheet is constantly gaining ice from snowfall and losing ice through outflow to the sea. West Antarctica is currently experiencing a net outflow of glacial ice, which will increase global sea level over time. A review of the scientific studies looking at data from 1992 to 2006 suggested that a net loss of around 50 gigatonnes of ice per year was a reasonable estimate (around 0.14 mm of sea level rise). Significant acceleration of outflow glaciers in the Amundsen Sea Embayment may have more than doubled this figure for 2006.

East Antarctica is a cold region with a ground base above sea level and occupies most of the continent. This area is dominated by small accumulations of snowfall which becomes ice and thus eventually seaward glacial flows. The mass balance of the East Antarctic Ice Sheet as a whole is thought to be slightly positive (lowering sea level) or near to balance. However, increased ice outflow has been suggested in some regions.

Effects of Global Warming

Some of Antarctica has been warming up, particularly strong warming has been noted on the Antarctic Peninsula. A study by Eric Steig published in 2009 noted for the first time that the continent-wide average surface temperature trend of Antarctica is slightly positive at >0.05 °C (0.09 °F) per decade from 1957 to 2006. This study also noted that West Antarctica has warmed by more than 0.1 °C (0.2 °F) per decade in the last 50 years, and this warming is strongest in winter and spring. This is partly offset by fall cooling in East Antarctica. There is evidence from one study that Antarctica is warming as a result of human carbon dioxide emissions. However, the small amount of surface warming in West Antarctica is not believed to be directly affecting the West Antarctic Ice Sheet's contribution to sea level. Instead the recent increases in glacier outflow are believed to be due to an inflow of warm water from the deep ocean, just off the continental shelf. The net contribution to sea level from the Antarctic Peninsula is more likely to be a direct result of the much greater atmospheric warming there.

In 2003 the Antarctic Peninsula's Larsen-B ice shelf collapsed. Between 28 February and 8 March 2008, about 570 square kilometres (220 sq mi) of ice from the Wilkins Ice Shelf on the southwest part of the peninsula collapsed, putting the remaining 15,000 km^2 (5,800 sq mi) of the ice shelf at risk. The ice was being held back by a "thread" of ice about 6 km (4 mi) wide, prior to its collapse on 5 April 2009. According to NASA, the most widespread Antarctic surface melting of the past 30 years occurred in 2005, when an area of ice comparable in size to California briefly melted and refroze; this may have resulted from temperatures rising to as high as 5 °C (41 °F).

Ozone Depletion

Each year a large area of low ozone concentration or "ozone hole" grows over Antarctica. This hole covers the whole continent and is at its largest in September. 2008 saw the longest lasting hole on record, which remained until the end of December. The hole was detected by scientists in 1985 and has tended to increase over the years of observation. The ozone hole is attributed to the emission of chlorofluorocarbons or CFCs into the atmosphere, which decompose the ozone into other gases. Some scientific studies suggest that ozone depletion may have a dominant role in the recent climate changes in Antarctica (and a wider area of the Southern Hemisphere). Ozone absorbs large amounts of ultraviolet radiation in the stratosphere. Ozone depletion over Antarctica can cause a cooling of around 6 °C in the local stratosphere.

This cooling has the effect of intensifying the westerly winds which flow around the continent (the polar vortex) and thus prevents outflow of the cold air near the South Pole. As a result, the continental mass of the East Antarctic ice sheet is held at lower temperatures, and the peripheral areas of Antarctica, especially the Antarctic Peninsula, are subject to higher temperatures, which promote accelerated melting. Recent models also suggest that the ozone depletion/enhanced polar vortex effect also accounts for the recent increase in sea-ice just offshore of the continent.

Antarctic Treaty System

The Antarctic Treaty and related agreements, collectively called the Antarctic Treaty System or ATS, regulate international relations with respect to Antarctica, Earth's only continent without a native human population. For the purposes of the treaty system, Antarctica is defined as all of the land and ice shelves south of 60°S latitude. The treaty, entering into force in 1961 and eventually signed by 47 countries, sets aside Antarctica as a scientific preserve, establishes freedom of scientific investigation and bans military activity on that continent. The treaty was the first arms control agreement established during the Cold War. The Antarctic Treaty Secretariat headquarters have been located in Buenos Aires, Argentina, since September 2004.

The Antarctic Treaty System

The main treaty was opened for signature on December 1, 1959, and officially entered into force on June 23, 1961. The original signatories were the 12 countries active in Antarctica during the International Geophysical Year (IGY) of 1957–58 and willing to accept a US invitation to the conference at which the treaty was negotiated. The twelve countries had significant interests in Antarctica at the time: Argentina, Australia, Belgium, Chile, France, Japan, New Zealand, Norway, South Africa, the Soviet Union, the United Kingdom and the United States. These countries had established over 50 Antarctic stations for the IGY. The treaty was a diplomatic expression of the operational and scientific cooperation that had been achieved "on the ice".

Articles of the Antarctic Treaty

- Article 1 – area to be used for peaceful purposes only; military activity, such as weapons testing is prohibited, but military personnel and equipment may be used for scientific research or any other peaceful purpose;
- Article 2 – freedom of scientific investigation and cooperation shall continue;
- Article 3 – free exchange of information and personnel in cooperation with the United Nations and other international agencies;
- Article 4 – the treaty does not recognize, dispute, nor establish territorial sovereignty claims; no new claims shall be asserted while the treaty is in force;
- Article 5 – prohibits nuclear explosions or disposal of radioactive wastes;
- Article 6 – includes under the treaty all land and ice shelves south of 60 degrees 00 minutes south;
- Article 7 – treaty-state observers have free access, including aerial observation, to any area and may inspect all stations, installations, and equipment; advance notice of all activities and of the introduction of military personnel must be given;
- Article 8 – allows for jurisdiction over observers and scientists by their own states;
- Article 9 – frequent consultative meetings take place among member nations;
- Article 10 – treaty states will discourage activities by any country in Antarctica that are contrary to the treaty;
- Article 11 – disputes to be settled peacefully by the parties concerned or, ultimately, by the International Court of Justice;

- Articles 12, 13, 14 – deal with upholding, interpreting, and amending the treaty among involved nations.

The main objective of the ATS is to ensure *in the interests of all humankind that Antarctica shall continue forever to be used exclusively for peaceful purposes and shall not become the scene or object of international discord*. The treaty forbids *any measures of a military nature*, but not the presence of military personnel.

Other Agreements

Other agreements — some 200 recommendations adopted at treaty consultative meetings and ratified by governments — include:

- Agreed Measures for the Conservation of Antarctic Fauna and Flora (1964) (entered into force in 1982)
- The Convention for the Conservation of Antarctic Seals (1972)
- The Convention for the Conservation of Antarctic Marine Living Resources (1980)
- The Convention on the Regulation of Antarctic Mineral Resource Activities (1988) (although it was signed in 1988, it was subsequently rejected and never entered into force)
- The Protocol on Environmental Protection to the Antarctic Treaty was signed October 4, 1991 and entered into force January 14, 1998; this agreement prevents development and provides for the protection of the Antarctic environment through five specific annexes on marine pollution, fauna and flora, environmental impact assessments, waste management, and protected areas. It prohibits all activities relating to mineral resources except scientific. A sixth annex — on liability arising from environmental emergencies — was adopted in 2005 but is yet to enter into force.

Meetings

The Antarctic Treaty System's yearly *Antarctic Treaty Consultative Meetings (ATCM)* are the international forum for the administration and management of the region. Only 28 of the 46 parties to the agreements have the right to participate in decision-making at these meetings, though the other 18 are still allowed to attend. The decision-making participants are the *Consultative Parties* and, in addition to the 12 original signatories, include 16 countries that have demonstrated their interest in Antarctica by carrying out substantial scientific activity there.

Members

Currently, there are 46 treaty member nations: 28 consultative and 18 acceding. Consultative (voting) members include the seven nations that claim portions of Antarctica as national territory. The 21 non-claimant nations either do not recognize the claims of others, or have not stated their positions.

Antarctic Treaty Secretariat

The Antarctic Treaty Secretariat was established in Buenos Aires, Argentina in September 2004 by the Antarctic Treaty Consultative Meeting (ATCM). Mr. Jan Huber (from the Netherlands) served as the first Executive Secretary for 5 years until August 31, 2009. He was succeeded on September 1, 2009 by Mr. Manfred Reinke (Germany), appointed for a 4 year term.

The tasks of the Antarctic Treaty Secretariat can be divided into the following areas:

- Supporting the annual Antarctic Treaty Consultative Meeting (ATCM) and the meeting of the Committee for Environmental Protection (CEP).
- Facilitating the exchange of information between the Parties required in the Treaty and the Environment Protocol.
- Collecting, storing, arranging and publishing the documents of the ATCM.
- Providing and disseminating public information about the Antarctic Treaty system and Antarctic activities.

There are 46 countries that own bases in Antarctica.

Legal System

Antarctica has no permanent population and hence no citizenship or government. All personnel present on Antarctica at any time are citizens or nationals of some sovereignty outside of Antarctica, as there is no Antarctic sovereignty. The majority of Antarctica is claimed by one or more countries, but most countries do not explicitly recognize those claims. The area on the mainland between 90 degrees west and 150 degrees west, combined with the interior of the Norwegian Sector (the extent of which has never been officially defined), is the only major land on Earth not claimed by any country.

Governments that are party to the Antarctic Treaty and its Protocol on Environmental Protection implement the articles of these agreements, and decisions taken under them, through national laws.

These laws generally apply only to their own citizens, wherever they are in Antarctica, and serve to enforce the consensus decisions of the consultative parties: about which activities are acceptable, which areas require permits to enter, what processes of environmental impact assessment must precede activities, and so on. The Antarctic Treaty is often considered to represent an example of the Common heritage of mankind principle.

Argentina

According to Argentine regulations, any crime committed within 50 kilometers of any Argentine base is to be judged in Ushuaia (as capital of Tierra del Fuego, Antarctica, and South Atlantic Islands). In the part of Argentine Antarctica that is also claimed by Chile and UK, the person to be judged can ask to be transferred there.

Australia

Since the designation of the Australian Antarctic Territory pre-dated the signing of the Antarctic Treaty, some of the complex suite of Australian laws that relate to Antarctica date from more than two decades before the Antarctic Treaty era. In terms of criminal law, the laws that apply to the Jervis Bay Territory (which follows the laws of the Australian Capital Territory) apply to the Australian Antarctic Territory. Key Australian legislation applying Antarctic Treaty System decisions include the *Antarctic Treaty Act 1960*, the *Antarctic Treaty (Environment Protection) Act 1980* and the *Antarctic Marine Living Resources Conservation Act 1981*.

United States

The law of the United States, including certain criminal offenses by or against U.S. nationals, such as murder, may apply to areas not under jurisdiction of other countries. To this end, the United States now stations special deputy U.S. Marshals in Antarctica to provide a law enforcement presence.

Some U.S. laws directly apply to Antarctica. For example, the Antarctic Conservation Act, Public Law 95-541, 16 U.S.C. section 2401 *et seq.*, provides civil and criminal penalties for the following activities, unless authorized by regulation or statute:

- the taking of native Antarctic mammals or birds
- the introduction into Antarctica of non-indigenous plants and animals
- entry into specially protected or scientific areas
- the discharge or disposal of pollutants into Antarctica or Antarctic waters
- the importation into the U.S. of certain items from Antarctica.

Violation of the Antarctic Conservation Act carries penalties of up to US$10,000 in fines and one year in prison. The Departments of the Treasury, Commerce, Transportation, and the Interior share enforcement responsibilities. The Act requires expeditions from the U.S. to Antarctica to notify, in advance, the Office of Oceans and Polar Affairs of the State Department, which reports such plans to other nations as required by the Antarctic Treaty. Further information is provided by the Office of Polar Programs of the National Science Foundation.

New Zealand

In 2006, the New Zealand police reported that jurisdictional issues prevented them issuing warrants for potential American witnesses who were reluctant to testify during the Christchurch Coroner's investigation into the death by poisoning of Australian astrophysicist Rodney Marks at the South Pole base in May 2000. Dr. Marks died while wintering over at the United States' Amundsen-Scott South Pole Station located at the geographic South Pole. Prior to autopsy, the death was attributed to natural causes by the National Science Foundation and the contractor administering the base. However, an autopsy in New Zealand revealed that Dr. Marks died from methanol poisoning. The New Zealand Police launched an investigation. In 2006, frustrated by lack of progress, the Christchurch Coroner said that it was unlikely that Dr. Marks ingested the methanol knowingly, although there is no certainty that he died as the direct result of the act of another person. During media interviews, the police detective in charge of the investigation criticized the National Science Foundation and contractor Raytheon for failing to co-operate with the investigation.

Arctic

The Arctic is the region around the Earth's North Pole, opposite the Antarctic region around the South Pole. The Arctic includes the Arctic Ocean (which overlies the North Pole) and parts of Canada, Greenland (a territory of Denmark), Russia, the United States (Alaska), Iceland, Norway, Sweden and Finland.

The Arctic region can be defined as the area north of the Arctic Circle (66° 33'N), which is the approximate limit of the midnight sun and the polar night. Alternatively, it can be defined as the region where the average temperature for the warmest month (July) is below 10 °C (50 °F); the northernmost tree line roughly follows the isotherm at the boundary of this region. Socially and politically, the Arctic region includes the northern territories of the eight Arctic states, although by natural science definitions much of this territory is considered subarctic. The Arctic region consists of a vast, ice-covered ocean (which is sometimes considered to be a northern arm of the Atlantic Ocean) surrounded by treeless permafrost. In recent years the extent of the sea ice has declined. Life in the Arctic includes organisms living in the ice, zooplankton and phytoplankton, fish and marine mammals, birds, land animals, plants, and human societies.

The Arctic region is a unique area among Earth's ecosystems. The cultures in the region and the Arctic indigenous peoples have adapted to its cold and extreme conditions.

Due to the poleward migration of the planet's isotherms (about 35 miles per decade during the past 30 years as a consequence of global warming), the Arctic region (as defined by tree line and temperature) is currently shrinking. Perhaps the most spectacular result of Arctic shrinkage is sea ice loss. There is a large variance in predictions of Arctic sea ice loss, with models showing near-complete to complete loss in September from 2040 to some time well beyond 2100. About half of the analyzed models show near-complete to complete sea ice loss in September by the year 2100.

Nature

Climate

The Arctic's climate is characterized by cold winters and cool summers. Precipitation mostly comes in the form of snow. The Arctic's annual precipitation is low, with most of the area receiving less than 50 centimetres (20 in). High winds often stir up snow, creating the illusion of continuous snowfall. Average winter temperatures can be as low as "40 °C ("40 °F), and the coldest recorded temperature is approximately "68 °C ("90 °F). Coastal Arctic climates are moderated by oceanic influences, having generally warmer temperatures and heavier snowfalls than the colder and drier interior areas. The Arctic is affected by current global warming, leading to Arctic shrinkage and Arctic methane release.

Plants

Arctic vegetation is composed of plants such as dwarf shrubs, graminoids, herbs, lichens and mosses, which all grow relatively close to the ground, forming tundra. As one moves northward, the amount of warmth available for plant growth decreases considerably. In the northernmost areas, plants are at their metabolic limits, and small differences in the total amount of summer warmth make large differences in the amount of energy available for maintenance, growth and reproduction. Colder summer temperatures cause the size, abundance, productivity and variety of plants to decrease. Trees cannot grow in the Arctic, but in its warmest parts, shrubs are common and can reach 2 m (6 ft 7 in) in height; sedges, mosses and lichens can form thick layers. In the coldest parts of the Arctic, much of the ground is bare; nonvascular plants such as lichens and mosses predominate, along with a few scattered grasses and forbs (like the arctic poppy).

Animals

Herbivores on the tundra include the Arctic hare, lemming, muskox, and caribou. They are preyed on by the Arctic fox and wolf. The polar bear is also a predator, though it prefers to hunt for marine life from the ice. There are also many birds and marine species endemic to the colder regions. Other land animals include wolverines, ermines, and arctic ground squirrels. Marine mammals include seals, walrus, and several species of cetacean—baleen whales and also narwhals, killer whales and belugas.

Natural Resources

The Arctic includes sizable natural resources (oil, gas, minerals, fresh water, forest—if the subarctic is included—and fish) to which modern technology and the economic opening up of Russia have given significant new opportunities. The interest of the tourism industry is also on the increase

The Arctic is one of the last and most extensive continuous wilderness areas in the world, and its significance in preserving biodiversity and genotypes is considerable. The increasing presence of humans fragments vital habitats. The Arctic is particularly susceptible to the abrasion of groundcover and to the disturbance of the rare reproduction places of the animals that are characteristic to the region. The Arctic also holds 1/5 of the Earth's water supply.

Paleo-history

During the Cretaceous, the Arctic still had seasonal snows, though only a light dusting and not enough to permanently hinder plant growth. Animals such as *Chasmosaurus*, *Hypacrosaurus*, *Troodon*, and *Edmontosaurus* may have all migrated north to take advantage of the summer growing season, and migrated south to warmer climes when the winter came. A similar situation may also have been found amongst dinosaurs that lived in Antarctic regions, such as *Muttaburrasaurus* of Australia.

Indigenous Population

The earliest inhabitants of North America's central and eastern Arctic are referred to as the Arctic small tool tradition

(AST) and existed circa 2500 BC. AST consisted of several subsets, including Independence culture and Pre-Dorset culture. The Dorset culture (Inuktitut: *Tuniit* or *Tunit*) refers to the next inhabitants of central and eastern Arctic.

The Dorset culture evolved because of technological and economic changes during the period of 1050-550 BC. With the exception of the Quebec/Labrador peninsula, the Dorset culture vanished around 1500 AD. Supported by genetic testing, evidence shows that Dorset culture survived in Aivilik, Southampton and Coats Islands, until the beginning of the 20th century.

Dorset/Thule culture transition dates around the 9th-10th centuries. Scientists theorize that there may have been cross-contact of the two cultures with sharing of technology, such as fashioning harpoon heads, or the Thule may have found Dorset remnants and adapted their ways with the predecessor culture. Others believe the Thule displaced the Dorset.

By 1300, the Inuit, present-day Arctic inhabitants and descendants of Thule culture, had settled west Greenland, and moved into east Greenland over the following century. Over time, the Inuit have migrated throughout the Arctic regions of Canada, Greenland, Russia and the United States.

Other Circumpolar North indigenous peoples include the Chukchi, Evenks, Inupiat, Khanty, Koryaks, Nenets, Sami, Yukaghir, and Yupik.

International Cooperation and Politics

The Arctic region is a focus of international political interest. International Arctic cooperation got underway on a broad scale well over ten years ago. The International Arctic Science Committee (IASC), hundreds of scientists and specialists of the Arctic Council, the Barents Euro-Arctic Council and its regional cooperation have compiled high quality information on the Arctic.

Territorial Claims

No country owns the geographic North Pole or the region of the Arctic Ocean surrounding it. The surrounding Arctic states that border the Arctic Ocean — Russia, Norway, the

United States, Canada and Denmark (via Greenland)—are limited to a 370 kilometre (200 nautical mile) economic zone around their coasts.

Upon ratification of the United Nations Convention on the Law of the Sea, a country has ten years to make claims to extend its 200 mile zone. Due to this, Norway (which ratified the convention in 1996), Russia (ratified in 1997), Canada (ratified in 2003) and Denmark (ratified in 2004) launched projects to establish claims that certain Arctic sectors should belong to their territories.

On August 2, 2007, two Russian bathyscaphes, MIR-1 and MIR-2, for the first time in history descended to the Arctic seabed beneath the North Pole and placed there a Russian flag made of rust-proof titanium alloy.

The mission was a scientific expedition, but the flag-placing raised concerns of a race for control of the Arctic's vast petroleum resources. (See 2007 Russian North Pole expedition.)

Foreign ministers and other officials representing Canada, Denmark, Norway, Russia, and the United States met in Ilulissat, Greenland on May 28, 2008 at the Arctic Ocean Conference and announced the Ilulissat Declaration.

Scientific Exploration

Since 1937, the whole Arctic region has been extensively explored by Soviet and Russian manned drifting ice stations. Between 1937 and 1991, 88 polar crews established and occupied scientific settlements on the drift ice and were carried thousands of kilometers by the ice flow.

Pollution

The Arctic is comparatively clean, although there are certain ecologically difficult localized pollution problems that present a serious threat to people's health living around these pollution sources. Due to the prevailing worldwide sea and air currents, the Arctic area is the fallout region for long-range transport pollutants, and in some places the concentrations exceed the levels of densely populated urban areas.

An example of this is the phenomenon of Arctic haze, which is commonly blamed on long-range pollutants. Another example is with the bioaccumulation of PCB's (polychlorinated biphenyls) in Arctic wildlife and people.

Climate Change

The Arctic is especially vulnerable to the effects of global warming, as has become apparent in the melting sea ice in recent years. Climate models predict much greater warming in the Arctic than the global average, resulting in significant international attention to the region.

In particular, there are concerns that Arctic shrinkage, a consequence of melting glaciers and other ice in Greenland, could soon contribute to a substantial rise in sea levels worldwide. Climate models give a range of predictions of Arctic sea ice loss, showing near-complete to complete loss in September anywhere from 2040 to some time well beyond 2100. About half of the analyzed models show near-complete to complete sea ice loss in September by the year 2100. More recently, the Catlin Arctic Survey concluded that summer ice loss would occur around 2029.

In September 2008, the extent of the summer Arctic ice cap was at a near-record low, only 9.01 percent greater than the record low in 2007, and 33.6 percent below the average extent of sea ice from 1979 to 2000.

The current Arctic shrinkage is leading to fears of Arctic methane release. Release of methane stored in permafrost could cause abrupt and severe global warming, as methane is a potent greenhouse gas. On millennial time-scales, decomposition of methane hydrates in the Arctic seabed could also amplify global warming. Previous methane release events have been linked to the great dying, a mass extinction event at the boundary of the Permian and Triassic, and the Paleocene-Eocene thermal maximum, in which temperatures abruptly increased.

Apart from concerns regarding the detrimental effects of warming in the Arctic, some potential opportunities have gained attention as well. The melting of the ice is making the Northwest passage, the shipping routes through the northernmost latitudes, more navigable, raising the possibility that the Arctic region will become a prime trade route. In addition, it is believed that the Arctic seabed may contain substantial oil fields which may become accessible if the ice covering them melts. These factors have led to recent international debates as to which nations can claim sovereignty or ownership over the waters of the Arctic.

The National Oceanic and Atmospheric Administration's Arctic Report Card presents annually updated, peer-reviewed information on recent observations of environmental conditions in the Arctic relative to historical records. In 2008, there continues to be widespread and, in some cases, dramatic evidence of an overall warming of the Arctic system.

10

Geography of Africa

Introduction

Africa is the world's second-largest and second most-populous continent, after Asia. At about 30.2 million km² (11.7 million sq mi) including adjacent islands, it covers 6% of the Earth's total surface area and 20.4% of the total land area. With a billion people in 61 territories, it accounts for about 14.72% of the World's human population.

The continent is surrounded by the Mediterranean Sea to the north, both the Suez Canal and the Red Sea along the Sinai Peninsula to the northeast, the Indian Ocean to the southeast, and the Atlantic Ocean to the west. The continent has 54 sovereign states, including Madagascar, various island groups, and the Sahrawi Arab Democratic Republic, a member state of the African Union whose statehood is disputed by Morocco.

Africa, particularly central eastern Africa, is widely regarded within the scientific community to be the origin of humans and the Hominidae clade (great apes), as evidenced by the discovery of the earliest hominids and their ancestors, as well as later ones that have been dated to around seven million years ago – including *Sahelanthropus tchadensis*, *Australopithecus Africans*, *A. afarensis*, *Homo erectus*, *H. habilis* and *H. Ergaster* – with the earliest *Homo sapiens* (modern human) found in Ethiopia being dated to circa 200,000 years ago.

Africa straddles the equator and encompasses numerous climate areas; it is the only continent to stretch from the northern temperate to southern temperate zones.

Etymology

Afri was the name of several peoples who dwelt in North Africa near Carthage. Their name is usually connected with Phoenician *afar*, "dust", but a 1981 theory has asserted that it stems from a Berber word *ifri* or Ifran meaning "cave", in reference to cave dwellers. Africa or Ifri or Afer is name of Banu Ifran from Algeria and Tripolitania (Berber Tribe of Yafran).

Under Roman rule, Carthage became the capital of Africa Province, which also included the coastal part of modern Libya. The Roman suffix "-ca" denotes "country or land". The later Muslim kingdom of Ifriqiya, modern-day Tunisia, also preserved a form of the name.

Other etymological theories that have been postulated for the ancient name "Africa":

- the 1st century Jewish historian Flavius Josephus (*Ant. 1.15*) asserted that it was named for Epher, grandson of Abraham according to Gen. 25:4, whose descendants, he claimed, had invaded Libya.
- Latin word *aprica* ("sunny") mentioned by Isidore of Seville in *Etymologiae* XIV.5.2.
- the Greek word *aphrike*, meaning "without cold." This was proposed by historian Leo Africans (1488–1554), who suggested the Greek word *phrike*, combined with the privative prefix "a-", thus indicating a land free of cold and horror.
- Massey, in 1881, derived an etymology from the Egyptian *af-rui-ka*, "to turn toward the opening of the Ka." The Ka is the energetic double of every person and "opening of the Ka" refers to a womb or birthplace. Africa would be, for the Egyptians, "the birthplace."
- yet another hypothesis was proposed by Michele Fruyt in *Revue de Philologie* 50, 1976: 221–238, linking the Latin word with *africus* 'south wind', which would be of Umbrian origin and mean originally 'rainy wind'.

The Irish female name *Aifric* is sometimes anglicised as *Africa*, but the given name is unrelated to the geonym.

History

At the beginning of the Mesozoic Era, Africa was joined with Earth's other continents in Pangaea. Africa shared the super continent's relatively uniform fauna which was dominated by theropods, prosauropods and primitive ornithischians by the close of the Triassic period. Late Triassic fossils are found throughout Africa, but are more common in the south than north. The boundary separating the Triassic and Jurassic marks the advent of an extinction event with global impact, although African strata from this time period have not been thoroughly studied.

Early Jurassic strata are distributed in a similar fashion to Late Triassic beds, with more common outcrops in the south and less common fossil beds which are predominated by tracks to the north. As the Jurassic proceeded, larger and more iconic groups of dinosaurs like sauropods and ornithopods proliferated in Africa. Middle Jurassic strata are neither well represented nor well studied in Africa. Late Jurassic strata are also poorly represented apart from the spectacular Tendaguru fauna in Tanzania. The Late Jurassic life of Tendaguru is very similar to that found in western North America's Morrison Formation.

Midway through the Mesozoic, about 150–160 million years ago, Madagascar separated from Africa, although it remained connected to India and the rest of the Gondwanan landmasses. Fossils from Madagascar include abelisaurs and titanosaurs.

Later into the Early Cretaceous epoch, the India-Madagascar landmass separated from the rest of Gondwana. By the Late Cretaceous, Madagascar and India had permanently split ways and continued until later reaching their modern configurations.

By contrast to Madagascar, mainland Africa was relatively stable in position throughout the Mesozoic. Despite the stable position, major changes occurred to its relation to other landmasses as the remains of Pangea continued to break apart. By the beginning of the Late Cretaceous epoch South America had split off from Africa, completing the southern half of the Atlantic Ocean. This event had a profound effect on global climate by altering ocean currents. During the Cretaceous, Africa was populated by allosauroids and spinosaurids, including the largest known carnivorous dinosaurs. Titanosaurs were significant herbivores in its ancient ecosystems. Cretaceous sites are more common than Jurassic ones, but are often unable to be dated radiometrically making it difficult to know their exact ages. Paleontologist Louis Jacobs, who spent time doing field work in Malawi, says that African beds are "in need of more field work" and will prove to be a "fertile ground ... for discovery."

Pre-history

Africa is considered by most paleoanthropologists to be the oldest inhabited territory on Earth, with the human species originating from the continent. During the middle of the twentieth century, anthropologists discovered many fossils and evidence of human occupation perhaps as early as 7 million years ago. Fossil remains of several species of early apelike humans thought to have evolved into modern man, such as *Australopithecus afarensis* (radiometrically dated to approximately 3.9–3.0 million years BC), *Paranthropus boisei* (c. 2.3–1.4 million years BC) and *Homo ergaster* (c. 1.9 million–600,000 years BC) have been discovered.

Throughout humanity's prehistory, Africa (like all other continents) had no nation states, and was instead inhabited by groups of hunter-gatherers such as the Khoi and San.

At the end of the Ice Ages, estimated to have been around 10,500 BC, the Sahara had again become a green fertile valley, and its African populations returned from the interior and coastal highlands in Sub-Saharan Africa.

However, the warming and drying climate meant that by 5000 BC the Sahara region was becoming increasingly dry and hostile. The population trekked out of the Sahara region towards the Nile Valley below the Second Cataract where they made permanent or semi-permanent settlements. A major climatic recession occurred, lessening the heavy and persistent rains in Central and Eastern Africa. Since this time dry conditions have prevailed in Eastern Africa, and increasingly during the last 200 years, in Ethiopia.

The domestication of cattle in Africa preceded agriculture and seems to have existed alongside hunter-gathering cultures. It is speculated that by 6000 BC cattle were already domesticated in North Africa. In the Sahara-Nile complex, people domesticated many animals including the donkey, and a small screw-horned goat which was common from Algeria to Nubia. In the year 4000 BC the climate of the Sahara started to become drier at an exceedingly fast pace.

This climate change caused lakes and rivers to shrink significantly and caused increasing desertification. This, in turn, decreased the amount of land conducive to settlements and helped to cause migrations of farming communities to the more tropical climate of West Africa.

By the first millennium BC iron working had been introduced in Northern Africa and quickly spread across the Sahara into the northern parts of sub-Saharan Africa and by 500 BC metalworking began to become commonplace in West Africa. Iron working was fully established by roughly 500 BC in many areas of East and West Africa, although other regions didn't begin iron working until the early centuries AD. Copper objects from Egypt, North Africa, Nubia and Ethiopia dating from around 500 BC have been excavated in West Africa, suggesting that trans-saharan trade networks had been established by this date.

Early Civilizations

At about 3300 BC, the historical record opens in Northern Africa with the rise of literacy in the Pharaonic civilisation of Ancient Egypt. One of the world's earliest and longest-lasting civilizations, the Egyptian state continued, with varying levels of influence over other areas, until 343 BC. Egyptian influence reached deep into modern-day Libya, north to Crete and Canaan, and south to the kingdoms of Aksum and Nubia.

An independent centre of civilisation with trading links to Phoenicia was established by Phoenicians from Tyre on the north-west African coast at Carthage.

European exploration of Africa began with Ancient Greeks and Romans. In 332 BC, Alexander the Great was welcomed as a liberator in Persian-occupied Egypt. He founded Alexandria in Egypt, which would become the prosperous capital of the Ptolemaic dynasty after his death. Following the conquest of North Africa's Mediterranean coastline by the Roman Empire, the area was integrated economically and culturally into the

Roman system. Roman settlement occurred in modern Tunisia and elsewhere along the coast. Christianity spread across these areas from Palestine via Egypt, also passing south, beyond the borders of the Roman world into Nubia and by at least the 6th century into Ethiopia.

In the early 7th century, the newly formed Arabian Islamic Caliphate expanded into Egypt, and then into North Africa. In a short while the local Berber elite had been integrated into Muslim Arab tribes. When the Ummayad capital Damascus fell in the eight century, the Islamic centre of the Mediterranean shifted from Syria to Qayrawan in North Africa. Islamic North Africa had become diverse, and a hub for mystics, scholars, jurists and philosophers. During the above mentioned period, Islam spread to sub-Saharan Africa, mainly through trade routes and migration.

9th–18th Century

Pre-colonial Africa possessed perhaps as many as 10,000 different states and polities characterised by many different sorts of political organisation and rule. These included small family groups of hunter-gatherers such as the San people of southern Africa; larger, more structured groups such as the family clan groupings of the Bantu-speaking people of central and southern Africa, heavily structured clan groups in the Horn of Africa, the large Sahelian kingdoms, and autonomous city-states and kingdoms such as those of the Yoruba and Igbo people (also misspelled as Ibo) in West Africa, and the Swahili coastal trading towns of East Africa.

By the 9th century AD a string of dynastic states, including the earliest Hausa states, stretched across the sub-saharan savannah from the western regions to central Sudan. The most powerful of these states were Ghana, Gao, and the Kanem-Bornu Empire. Ghana declined in the 11th century but was succeeded by the Mali Empire which consolidated much of western Sudan in the 13th century. Kanem accepted Islam in the 11th century.

In the forested regions of the West African coast, independent kingdoms grew up with little influence from the Muslim north. The Kingdom of Nri of the Igbo was established around the 9th century and was one of the first. It is also one of the oldest Kingdom in modern day Nigeria and was ruled by the Eze Nri. The Nri kingdom is famous for its elaborate bronzes, found at the town of Igbo Ukwu. The bronzes have been dated from as far back as the 9th century.

The Ife, historically the first of these Yoruba city-states or kingdoms, established government under a priestly oba (ruler), (oba means 'king' or 'ruler' in the Yoruba language), called the *Ooni of Ife*. Ife was noted as a major religious and cultural centre in Africa, and for its unique naturalistic tradition of bronze sculpture. The Ife model of government was adapted at Oyo, where its obas or kings, called the *Alaafins of Oyo* once controlled a large number of other Yoruba and non Yoruba city states and Kingdoms, the Fon *Kingdom of Dahomey* was one of the non Yoruba domains under Oyo control.

The Almoravids, was a Berber dynasty from the Sahara that spread over a wide area of northwestern Africa and the Iberian peninsula during the 11th century. The Banu Hilal and Banu Maqil were a collection of Arab Bedouin tribes from the Arabian peninsula who migrated westwards via Egypt between the 11th and 13th centuries. Their migration resulted in the fusion of the Arabs and Berbers, where the locals were Arabized, and Arab culture absorbed elements of the local culture, under the unifying framework of Islam.

Following the breakup of Mali a local leader named Sonni Ali (1464–1492) founded the Songhai Empire in the region of middle Niger and the western Sudan and took control of the trans-Saharan trade. Sonni Ali seized Timbuktu in 1468 and Jenne in 1473, building his regime on trade revenues and the cooperation of Muslim merchants. His successor Askia Mohammad I (1493–1528) made Islam the official religion, built mosques, and brought Muslim scholars, including al-Maghili, the founder of an important tradition of Sudanic African Muslim scholarship, to Gao. By the 11th century some Hausa states – such as Kano, jigawa, Katsina, and Gobir – had developed into walled towns engaging in trade, servicing caravans, and the manufacture of goods. Until the 15th century these small states were on the periphery of the major Sudanic empires of the era, paying tribute to Songhai to the west and Kanem-Borno to the east.

Height of Slave Trade

Slavery had long been practiced in Africa. Between the seventh and twentieth centuries, Arab slave trade (also known as slavery in the East) took 18 million slaves from Africa via trans-Saharan and Indian Ocean routes. Between the fifteenth and the nineteenth centuries, the Atlantic slave trade took 7–12 million slaves to the New World.

In West Africa, the decline of the Atlantic slave trade in the 1820s caused dramatic economic shifts in local polities. The gradual decline of slave-trading, prompted by a lack of demand for slaves in the New World, increasing anti-slavery legislation in Europe and America, and the British Royal Navy's increasing presence off the West African coast, obliged African states to adopt new economies. Between 1808 and 1860, the British West Africa Squadron seized approximately 1,600 slave ships and freed 150,000 Africans who were aboard.

Action was also taken against African leaders who refused to agree to British treaties to outlaw the trade, for example against "the usurping King of Lagos", deposed in 1851. Anti-slavery treaties were signed with over 50 African rulers. The largest powers of West Africa: the Asante Confederacy, the Kingdom of Dahomey, and the Oyo Empire, adopted different ways of adapting to the shift. Asante and Dahomey concentrated on the development of "legitimate commerce" in the form of palm oil, cocoa, timber and gold, forming the bedrock of West Africa's modern export trade. The Oyo Empire, unable to adapt, collapsed into civil wars.

Colonialism and the "Scramble for Africa"

In the late nineteenth century, the European imperial powers engaged in a major territorial scramble and occupied most of the continent, creating many colonial territories, and leaving only two fully independent states: Ethiopia (known to Europeans as "Abyssinia"), and Liberia. Egypt and Sudan were never incorporated into any European colonial empire, however, after the British occupation of 1882, the country was effectively under British administration until 1922. Imperial rule by Europeans would continue until after the conclusion of World War II, when almost all remaining colonial territories gradually obtained formal independence.

Independence movements in Africa gained momentum following World War II, which left the major European powers weakened. In 1951, Libya, a former Italian colony, gained independence. In 1956, Tunisia and Morocco won their independence from France. Ghana followed suit the next year, becoming the first of the sub-Saharan colonies to be freed. Most of the rest of the continent became independent over the next decade, most often through relatively peaceful means, though in some countries, notably Algeria, it came only after a violent struggle.

Portugal's overseas presence in Sub-Saharan Africa (most notably in Angola, Cape Verde, Mozambique, Guinea and Sao Tome and Principe) lasted from the 16th century to 1975, after the Estado Novo regime was overthrown in a military coup in Lisbon. Zimbabwe won its independence from the United Kingdom in 1980 after a bitter guerrilla war between black nationalists and the white minority Rhodesian government of Ian Smith. Although South Africa was one of the first African countries to gain independence, the state remained under the control of the country's white minority through a system of racial segregation known as apartheid until 1994.

Post-colonial Africa

Today, Africa contains 54 sovereign countries, most of which still have the borders drawn during the era of European colonialism. Since colonialism, African states have frequently been hampered by instability, corruption, violence, and authoritarianism. The vast majority of African states are republics that operate under some form of the presidential system of rule. However, few of them have been able to sustain democratic governments on a permanent basis, and many have instead cycled through a series of coups, producing military dictatorships. A number of Africa's post-colonial political leaders were military generals who were poorly educated and ignorant on matters of governance. Great instability, however, was mainly the result of marginalization of ethnic groups, and graft under these leaders. For political gain, many leaders fanned ethnic conflicts that had been exacerbated, or even created, by colonial rule. In many countries, the military was perceived as being the only group that could effectively maintain order, and it ruled many nations in Africa during the 1970s and early 1980s. During the period from the early 1960s to the late 1980s, Africa had more than 70 coups and 13 presidential assassinations. Border and territorial disputes were also common, with the European-imposed borders of many nations being widely contested through armed conflicts.

Cold War conflicts between the United States and the Soviet Union, as well as the policies of the International Monetary Fund, also played a role in instability. When a country became independent for the first time, it was often expected to align with one of the two superpowers. Many countries in Northern Africa received Soviet military aid, while many in Central and Southern Africa were supported by the United States, France or both. The 1970s saw an escalation, as newly independent Angola and Mozambique aligned themselves with the Soviet Union, and the West and South Africa sought to contain Soviet influence by funding insurgency movements. There was a major famine in Ethiopia, when hundreds of thousands of people starved. Some claimed that Marxist/Soviet policies made the situation worse.

The most devastating military conflict in modern independent Africa has been the Second Congo War. By 2008, this conflict and its aftermath had killed 5.4 million people. Since 2003 there has been an ongoing conflict in Darfur which has become a humanitarian disaster. AIDS has also been a prevalent issue in post-colonial Africa.

Geography

Africa is a continent comprising 62 political territories (including 59 countries), representing the largest of the great southward projections from the main mass of Earth's surface. It includes, within its remarkably regular outline, an area of 30,368,609 km (11,725,385 sq mi), including adjacent islands. Africa is in the field of Geography usually divided into 4 main following parts: a) Northern Africa -predominantly Muslim, Arabic with dry and arid/semiarid climate conditions; mainly French speaking b) Western Africa -mixed Christian/Muslim population with higher level of precipitation in the average than the North Africa c) Eastern Africa -mixed Christian/Muslim population, very attractive for tourism (especially Kenya,

Tanzania, Burundi and Mozambique) d) Southern Africa - majority of the population Christian with The Republic of South Africa being the centre of tourism. Mainly English spoken. Separated from Europe by the Mediterranean Sea and from much of Asia by the Red Sea, Africa is joined to Asia at its northeast extremity by the Isthmus of Suez (which is transected by the Suez Canal), 130 km (81 mi) wide. For geopolitical purposes, the Sinai Peninsula of Egypt – east of the Suez Canal – is often considered part of Africa. From the most northerly point, Ras ben Sakka in Tunisia, in 37°212 N, to the most southerly point, Cape Agulhas in South Africa, 34°512 153 S, is a distance approximately of 8,000 km (5,000 mi); from Cape Verde, 17°332 223 W, the western most point, to Ras Hafun in Somalia, 51°272 523 E, the most easterly projection, is a distance (also approximately) of 7,400 km (4,600 mi). The length of coast-line is 26,000 km² (16,000 mi) and the absence of deep indentations of the shore is shown by the fact that Europe, which covers only 10,400,000 km² (4,000,000 sq mi), has a coastline of 32,000 km (20,000 mi).

The main structural lines of the continent show both the east-to-west direction characteristic, at least in the eastern hemisphere, of the more northern parts of the world, and the north-to-south direction seen in the southern peninsulas. Africa is thus composed of two segments at right angles, the northern running from east to west, the southern from north to south, the subordinate lines corresponding in the main to these two directions.

Plateau Region

The high southern and eastern plateaus, rarely falling below 600 m (2,000 ft), have a mean elevation of about 1,000 m (3,300 ft). The South African plateau, as far as about 12° S, is bounded east, west and south by bands of high ground which fall steeply to the coasts. On this account South Africa has a general resemblance to an inverted saucer. Due south the plateau rim is formed by three parallel steps with level ground between them. The largest of these level areas, the Great Karoo, is a dry, barren region, and a large tract of the plateau proper is of a still more arid character and is known as the Kalahari Desert.

The South African plateau is connected towards East African plateau, with probably a slightly greater average elevation, and marked by some distinct features. It is formed by a widening out of the eastern axis of high ground, which becomes subdivided into a number of zones running north and south and consisting in turn of ranges, tablelands and depressions. The most striking feature is the existence of two great lines of depression, due largely to the subsidence of whole segments of the Earth's crust, the lowest parts of which are occupied by vast lakes. Towards the south the two lines converge and give place to one great valley (occupied by Lake Nyasa), the southern part of which is less distinctly due to rifting and subsidence than the rest of the system.

Farther north the western depression, known as the Great Rift Valley is occupied for more than half its length by water, forming the Great Lakes of Tanganyika, Kivu, Lake Edward and Lake Albert, the first-named over 400 miles (600 km) long and the longest freshwater lake in the world. Associated with these great valleys are a number of volcanic peaks, the greatest of which occur on a meridional line east of the eastern trough. The eastern depression, known as the East African trough or rift-valley, contains much smaller lakes, many of them brackish and without outlet, the only one comparable to those of the western trough being Lake Turkana or Basso Norok.

At no great distance east of this rift-valley is Mount Kilimanjaro - with its two peaks Kibo and Mawenzi, the latter being 5,889 m (19,321 ft), and the culminating point of the whole continent — and Mount Kenya, which is 5,184 m (17,008 ft). Hardly less important is the Ruwenzori Range, over 5,060 m (16,600 ft), which lies east of the western trough. Other volcanic peaks rise from the floor of the valleys, some of the Kirunga (Mfumbiro) group, north of Lake Kivu, being still partially active.

The third division of the higher region of Africa is formed by the Ethiopian Highlands, a rugged mass of mountains forming the largest continuous area of its altitude in the whole continent, little of its surface falling below 1,500 m (4,900 ft), while the summits reach heights of 4600 m to 4900 m (15,000 to 16,000 ft). This block of country lies just west of the line of the great East African Trough, the northern continuation of which passes along its eastern escarpment as it runs up to join the Red Sea. There is, however, in the centre a circular basin occupied by Lake Tsana.

Both in the east and west of the continent the bordering highlands are continued as strips of plateau parallel to the coast, the Ethiopian mountains being continued northwards along the Red Sea coast by a series of ridges reaching in places a height of 2,000 m (6,600 ft). In the west the zone of high land is broader but somewhat lower. The most mountainous districts lie inland from the head of the Gulf of Guinea (Adamawa, etc.), where heights of 1800 m to 2400 m (6000 to 8000 ft) are reached. Exactly at the head of the gulf the great peak of the Cameroon, on a line of volcanic action continued by the islands to the south-west, has a height of 4,075 m (13,369 ft), while Clarence Peak, in Fernando Po, the first of the line of islands, rises to over 2,700 m (8,900 ft). Towards the extreme west the Futa Jallon highlands form an important diverging point of rivers, but beyond this, as far as the Atlas chain, the elevated rim of the continent is almost wanting.

Plains

The area between the east and west coast highlands, which north of 17° N is mainly desert, is divided into separate basins by other bands of high ground, one of which runs nearly centrally through North Africa in a line corresponding roughly with the curved axis of the continent as a whole. The best marked of the basins so formed (the Congo basin) occupies a circular area bisected by the equator, once probably the site of an inland sea.

Running along the south of desert is the plains region known as the Sahel. The arid region, the Sahara — the largest desert in the world, covering 9,000,000 km^2 (3,500,000 sq mi) — extends from the Atlantic to the Red Sea. Though generally of slight elevation it contains mountain ranges with peaks rising to 2,400 m (7,900 ft) Bordered N.W. by the Atlas range, to the northeast a rocky plateau separates it from the Mediterranean; this plateau gives place at the extreme east to the delta of the Nile. That river pierces the desert without modifying its character. The Atlas range, the north-westerly part of the continent, between its seaward and landward heights encloses elevated steppes in places 160 km (99 mi) broad. From the inner slopes of the plateau numerous wadis take a direction towards the Sahara. The greater part of that now desert region is, indeed, furrowed by old water-channels.

Hydrology

From the outer margin of the African plateaus, a large number of streams run to the sea with comparatively short courses, while the larger rivers flow for long distances on the interior highlands, before breaking through the outer ranges. The main drainage of the continent is to the north and west, or towards the basin of the Atlantic Ocean.

To the main African rivers belong: Nile (the longest river of Africa), Congo (river with the highest water discharge on the continent) and the Niger, which flows half of it's length through the arid areas. The largest lake are the following: Lake Victoria (Lake Ukerewe), Lake Chad, in the centre of the continent, Lake Tanganika laying between the Democratic Republic of Congo, Burundi, Tanzania & Zambia.

There is also a considerably large lake Malawi stretching along the eastern border of 1 of the poorest countries in the world -Malawi etc.. There are also numerous waterdams throughout the continent: Kariba on the river of Zambezi, Asuan in Egypt on the river of Nile and the biggest dam of the continent laying completely in The republic of Ghana is called Akosombo on the Volta river (Fobil 2003). The high lake plateau of East Africa contains the headwaters of both the Nile and the Congo.

The upper Nile receives its chief supplies from the mountainous region adjoining the Central African trough in the neighbourhood of the equator. From there, streams pour eastward into Lake Victoria, the largest African lake (covering over 26,000 square m.), and to the west and north into Lake Edward and Lake Albert. To the latter of these, the effluents of the other two lakes add their waters. Issuing from there, the Nile flows northward, and between the latitudes of 7 and 10 degrees N. it traverses a vast marshy level, where its course is liable to being blocked by floating vegetation. After receiving the Bahr-el-Ghazal from the west and the Sobat, Blue Nile and Atbara from the Ethiopian highlands (the chief gathering ground of the flood-water), it separates the great desert with its fertile watershed, and enters the Mediterranean at a vast delta. The most remote head-stream of the Congo is the Chambezi, which flows southwest into the marshy Lake Bangweulu. From this lake issues the Congo, known in its upper course by various names.

Flowing first south, it afterwards turns north through Lake Mweru and descends to the forest-clad basin of west equatorial Africa. Traversing this in a majestic northward curve, and receiving vast supplies of water from many great tributaries, it finally turns southwest and cuts a way to the Atlantic Ocean through the western highlands.

North of the Congo basin, and separated from it by a broad undulation of the surface, is the basin of Lake Chad - a flat-shored, shallow lake filled principally by the Chari coming from the southeast.

West of this is the basin of the Niger, the third major river of Africa. With its principal source in the far west, it reverses the direction of flow exhibited by the Nile and Congo, and ultimately flows into the Atlantic — a fact that eluded European geographers for many centuries. An important branch, however — the Benue - flows from the southeast.

These four river-basins occupy the greater part of the lower plateaus of North and West Africa — the remainder consisting of arid regions watered only by intermittent streams that do not reach the sea.

Of the remaining rivers of the Atlantic basin, the Orange, in the extreme south, brings the drainage from the Drakensberg on the opposite side of the continent, while the Kunene, Kwanza, Ogowe and Sanaga drain the west coastal highlands of the southern limb; the Volta, Komoe, Bandama, Gambia and Senegal the highlands of the western limb. North of the Senegal, for over 1000 miles (1600 km) of coast, the arid region reaches to the Atlantic. Farther north are the streams, with comparatively short courses, reaching the Atlantic and Mediterranean from the Atlas mountains.

Of the rivers flowing to the Indian Ocean, the only one draining any large part of the interior plateaus is the Zambezi, whose western branches rise in the western coastal highlands. The main stream has its rise in 11°212 33 S 24°222 E, at an elevation of 5,000 ft (1,500 m) It flows to the west and south for a considerable distance before turning eastward. All the largest tributaries, including the Shire, the outflow of Lake Nyasa, flow down the southern slopes of the band of high ground stretching across the continent from 10 deg. to 12 deg. S. In the southwest, the Zambezi system interlaces with that of the Taukhe (or Tioghe), from which it at times receives surplus water.

The rest of the water of the Taukhe, known in its middle course as the Okavango, is lost in a system of swamps and saltpans that was formerly centred in Lake Ngami, now dried up.

Farther south, the Limpopo drains a portion of the interior plateau, but breaks through the bounding highlands on the side of the continent nearest its source. The Rovuma, Rufiji, Tana, Jubba and Webi Shebeli principally drain the outer slopes of the East African highlands, the last of these losing itself in the sands in proximity to the sea. Another large stream, the Hawash, rising in the Ethiopian mountains, is lost in a saline depression near the Gulf of Aden.

Lastly, between the basins of the Atlantic and Indian Oceans, there is an area of inland drainage along the centre of the East African plateau, directed chiefly into the lakes in the great rift valley. The largest river is the Omo, which, fed by the rains of the Ethiopian highlands, carries down a large body of water into Lake Rudolf.

The rivers of Africa are generally obstructed either by bars at their mouths, or by cataracts at no great distance upstream. But when these obstacles have been overcome, the rivers and lakes afford a vast network of navigable waters. The calculation of the areas of African drainage systems, made by Dr. A. Bludau (Petermanns Mitteilungen, 43, 1897, pp. 184–186) yields the following general results:

	mi^2	Mm^2
Basin of the Atlantic	4,070,000	10.541
Basin of the Mediterranean	1,680,000	4.351
Basin of the Indian Ocean	2,086,000	5.403
Inland drainage area	3,452,000	8.941

The areas of individual river basins are:

	mi²	Mm²
Congo, length over 3000 mi (4800 km)	1,425,000	3.691
Nile, length fully 4000 mi (6500 km)	1,082,000	2.802
Niger, length about 2600 mi (4200 km)	808,000	2.093
Zambezi, length about 2000 mi (3200 km)	513,500	1.330
Lake Chad	394,000	1.020
Orange, length about 1300 mi (2100 km)	370,505	0.9596
Orange (actual drainage area)	172,500	0.447&0.467

The area of the Congo basin is greater than that of any other river except the Amazon, while the African inland drainage area is greater than that of any continent but Asia, where the corresponding area is 4,000,000 square miles (10 Mm²).

The principal African lakes have been mentioned in the description of the East African plateau, but some of the phenomena connected with them may be spoken of more particularly here. As a rule, the lakes found within the great rift-valleys have steep sides and are very deep. This is the case with the two largest of the type, Tanganyika and Nyasa, the latter with depths of 430 fathoms (790 m).

Others, however, are shallow, and hardly reach the steep sides of the valleys in the dry season. Such are Lake Rukwa, in a subsidiary depression north of Nyasa, and Eiassi and Manyara in the system of the eastern rift-valley. Lakes of the broad type are of moderate depth, the deepest sounding in Lake Victoria being under 50 fathoms (90 m).

Besides the East African lakes, the principal are: - Lake Chad, in the northern inland watershed; Bangweulu and Mweru, traversed by the head-stream of the Congo; and Lake Mai-Ndombe and Ntomba (Mantumba), within the great bend of that river. All, except possibly Mweru, are more or less shallow, and Lake Chad appears to be drying up.

Divergent opinions have been held as to the mode of origin of the East African lakes, especially Tanganyika, which some geologists have considered to represent an old arm of the sea, dating from a time when the whole central Congo basin was under water; others holding that the lake water has accumulated in a depression caused by subsidence. The former view is based on the existence in the lake of organisms of a decidedly marine type. They include jellyfish, molluscs, prawns, crabs, etc.

Islands

With the exception of Madagascar the African islands are small. Madagascar, with an area of 229,820 square miles (595,200 km^2), is, after Greenland, New Guinea and Borneo, the fourth largest island on the Earth. It lies off the S.E. coast of the continent, from which it is separated by the deep Mozambique channel, 250 miles (400 km) wide at its narrowest point. Madagascar in its general structure, as in flora and fauna, forms a connecting link between Africa and southern Asia. East of Madagascar are the small islands of Mauritius and Reunion. There are also islands in the Gulf of Guinea on which lies the Republic of Sao Tome and Principe (islands of Sao Tome and Principe). Part of the Republic of Equatorial Guinea is laying on the island of Bioko (with the capital Malabo and the town of Lubu) and the island of Annobon. Socotra lies E.N.E. of Cape Guardafui. Off the north-west coast are the Canary and Cape Verde archipelagoes. which, like some small islands in the Gulf of Guinea, are of volcanic origin.

Climate and Health

Lying almost entirely within the tropics, and equally to north and south of the equator, Africa does not show excessive variations of temperature.

Great heat is experienced in the lower plains and desert regions of North Africa, removed by the great width of the continent from the influence of the ocean, and here, too, the contrast between day and night, and between summer and winter, is greatest. (The rarity of the air and the great radiation during the night cause the temperature in the Sahara to fall occasionally to freezing point.)

Farther south, the heat is to some extent modified by the moisture brought from the ocean, and by the greater elevation of a large part of the surface, especially in East Africa, where the range of temperature is wider than in the Congo basin or on the Guinea coast.

In the extreme north and south the climate is a warm temperate one, the northern countries being on the whole hotter and drier than those in the southern zone; the south of the continent being narrower than the north, the influence of the surrounding ocean is more felt.

The most important climatic differences are due to variations in the amount of rainfall. The wide heated plains of the Sahara, and in a lesser degree the corresponding zone of the Kalahari in the south, have an exceedingly scanty rainfall, the winds which blow over them from the ocean losing part of their moisture as they pass over the outer highlands, and becoming constantly drier owing to the heating effects of the burning soil of the interior; while the scarcity of mountain ranges in the more central parts likewise tends to prevent condensation. In the inter-tropical zone of summer precipitation, the rainfall is greatest when the sun is vertical or soon after. It is therefore greatest of all near the equator, where the sun is twice vertical, and less in the direction of both tropics.

The rainfall zones are, however, somewhat deflected from a due west-to-east direction, the drier northern conditions extending southwards along the east coast, and those of the south northwards along the west. Within the equatorial zone certain areas, especially on the shores of the Gulf of Guinea and in the upper Nile basin, have an intensified rainfall, but this rarely approaches that of the rainiest regions of the world. The rainiest district in all Africa is a strip of coastland west of Mount Cameroon, where there is a mean annual rainfall of about 390 in (9,906 mm) as compared with a mean of 458 in (11,600 mm) at Cherrapunji, in Meghalaya, India.

The two distinct rainy seasons of the equatorial zone, where the sun is vertical at half-yearly intervals, become gradually merged into one in the direction of the tropics, where the sun is overhead but once. Snow falls on all the higher mountain ranges, and on the highest the climate is thoroughly Alpine.

The countries bordering the Sahara are much exposed to a very dry wind, full of fine particles of sand, blowing from the desert towards the sea. Known in Egypt as the khamsin, on the Mediterranean as the sirocco, it is called on the Guinea coast the harmattan. This wind is not invariably hot; its great dryness causes so much evaporation that cold is not infrequently the result. Similar dry winds blow from the Kalahari Desert in the south. On the eastern coast the monsoons of the Indian Ocean are regularly felt, and on the southeast hurricanes are occasionally experienced.

Extreme Points

This is a list of the extreme points of Africa, the points that are farther north, south, east or west than any other location on the continent.

Africa;

- Northernmost Point — Galite Islands, Tunisia (37°32'N)
- Southern most Point — Cape Agulhas, South Africa (34°51'15"S)
- Western most Point — Santo Antao, Cape Verde Islands (25°25'W)
- Easternmost Point — Rodrigues, Mauritius (63°30'E)
- African pole of inaccessibility is close to the border of Central African Republic, Sudan and Congo, near the town Obo.

Africa (mainland);

- Northernmost Point — Ras ben Sakka (Ra's al Abyad) (Cape Blanc), Tunisia
- Southern most Point — Cape Agulhas, South Africa
- Western most Point — Pointe des Almadies, Cap Vert Peninsula, Senegal (17°33'22"W)
- Easternmost Point — Ras Hafun (Raas Xaafuun), Somalia (51°27'52"E).

The highest point in Africa is Mount Kilimanjaro, 5,891.8 m (19,330 ft) in Tanzania. The lowest point is Lake Asal, 153 m (502 ft) below sea level, in Djibouti.

Africa is the largest of the three great southward projections from the largest landmass of the Earth. Separated from Europe by the Mediterranean Sea, it is joined to Asia at its northeast extremity by the Isthmus of Suez (transected by the Suez Canal), 163 km (101 miles) wide. (Geopolitically, Egypt's Sinai Peninsula east of the Suez Canal is often considered part of Africa, as well.)

From the most northerly point, Ras ben Sakka in Tunisia (37°21' N), to the most southerly point, Cape Agulhas in South

Africa (34°51'15" S), is a distance of approximately 8,000 km (5,000 miles); from Cape Verde, 17°33'22" W, the western most point, to Ras Hafun in Somalia, 51°27'52" E, the most easterly projection, is a distance of approximately 7,400 km (4,600 miles). The coastline is 26,000 km (16,100 miles) long, and the absence of deep indentations of the shore is illustrated by the fact that Europe, which covers only 10,400,000 km^2 (4,010,000 square miles) – about a third of the surface of Africa – has a coastline of 32,000 km (19,800 miles).

Africa's largest country is Sudan, and its smallest country is the Seychelles, an archipelago off the east coast. The smallest nation on the continental mainland is The Gambia.

According to the ancient Romans, Africa lay to the west of Egypt, while "Asia" was used to refer to Anatolia and lands to the east. A definite line was drawn between the two continents by the geographer Ptolemy (85–165 AD), indicating Alexandria along the Prime Meridian and making the isthmus of Suez and the Red Sea the boundary between Asia and Africa. As Europeans came to understand the real extent of the continent, the idea of *Africa* expanded with their knowledge.

Geologically, Africa includes the Arabian Peninsula; the Zagros Mountains of Iran and the Anatolian Plateau of Turkey mark where the African Plate collided with Eurasia. The Afrotropic ecozone and the Saharo-Arabian desert to its north unite the region biogeographically, and the Afro-Asiatic language family unites the north linguistically.

Climate

The climate of Africa ranges from tropical to subarctic on its highest peaks. Its northern half is primarily desert or arid, while its central and southern areas contain both savanna plains and very dense jungle (rainforest) regions. In between, there is a convergence where vegetation patterns such as sahel, and steppe dominate.

Fauna

Africa boasts perhaps the world's largest combination of density and "range of freedom" of wild animal populations and diversity, with wild populations of large carnivores (such as lions, hyenas, and cheetahs) and herbivores (such as buffalo, deer, elephants, camels, and giraffes) ranging freely on primarily open non-private plains. It is also home to a variety of "jungle" animals including snakes and primates and aquatic life such as crocodiles and amphibians. Africa also has the largest number of megaphone species, as it was least affected by the extinction of the Pleistocene megaphone.

Ecology

Deforestation is affecting Africa at twice the world rate, according to the United Nations Environment Programme (UNEP). Some sources claim that deforestation has already wiped out roughly 90% of West Africa's original forests. Since the arrival of humans 2000 years ago, Madagascar has lost more than 90% of its original forest. About 65% of Africa's agricultural land suffers from soil degradation.

Politics

The African Union (AU) is a 53 member federation consisting of all of Africa's states except Morocco. The union was formed, with Addis Ababa as its headquarters, on 26 June 2001. In July 2004, the African Union's Pan-African Parliament (PAP) was relocated to Midrand, in South Africa, but the African Commission on Human and Peoples' Rights remained in Addis Ababa. There is a policy in effect to decentralize the African Federation's institutions so that they are shared by all the states.

The African Union, not to be confused with the AU Commission, is formed by the Constitutive Act of the African Union, which aims to transform the African Economic Community, a federated commonwealth, into a state under established international conventions. The African Union has a parliamentary government, known as the African Union Government, consisting of legislative, judicial and executive organs. It is led by the African Union President and Head of State, who is also the President of the Pan African Parliament. A person becomes AU President by being elected to the PAP, and subsequently gaining majority support in the PAP. 1 The powers and authority of the President of the African Parliament derive from the Constitutive Act and the Protocol of the Pan African Parliament, as well as the inheritance of presidential authority stipulated by African treaties and by international treaties, including those subordinating the Secretary General of the OAU Secretariat (AU Commission) to the PAP. The government of the AU consists of all-union (federal), regional, state, and municipal authorities, as well as hundreds of institutions, that together manage the day-to-day affairs of the institution.

There are clear signs of increased networking among African organisations and states. In the civil war in the Democratic Republic of the Congo (former Zaire), rather than rich, non-African countries intervening, neighbouring African countries became involved. Since the conflict began in 1998, the estimated death toll has reached 5 million.

Political associations such as the African Union offer hope for greater co-operation and peace between the continent's many countries. Extensive human rights abuses still occur in several parts of Africa, often under the oversight of the state. Most of such violations occur for political reasons, often as a side effect of civil war. Countries where major human rights violations have been reported in recent times include the Democratic Republic of the Congo, Sierra Leone, Liberia, Sudan, Zimbabwe, and Cote d'Ivoire.

Economy

The economy of Africa is a of the trade, industry, and resources of the people of Africa. As of 2006, approximately 922 million people were living in 54 different countries. Africa is the world's poorest inhabited continent. Though parts of the continent have made significant gains over the last few years, of the 175 countries reviewed in the United Nations' Human Development Report 2003, 25 African nations ranked lowest amongst the nations of the world. This is partly due to its turbulent history. The decolonization of Africa was fraught with instability aggravated by cold war conflict. Since mid-20th century, the Cold War and increased corruption and despotism have also contributed to Africa's poor economy.

The biggest contrast in terms of development has been between Africa and the economy of Europe. The African Economic Outlook report specifically mentions that Africa's trade with China has multiplied by 10 since 2001, reaching over USD 100 billion in 2008. The economies of China and India have grown rapidly, while Latin America has also experienced moderate growth, lifting millions above subsistence living. By contrast, much of Africa has stagnated and even regressed in terms of foreign trade, investment, per capita income, and other economic growth measures. Poverty has had widespread effects, including low life expectancy, violence, and instability, which in turn have perpetuated the continent's growth problems. Over the decades, there have been many unsuccessful attempts to improve the economies of individual African countries. However, recent data suggest some parts of the continent are experiencing faster growth. The World Bank reports the economy of Sub-Saharan African countries grew at rates that match global rates. The economies of the fastest growing African nations experienced growth significantly above the global average rates. The top nations in 2007 include Mauritania with growth at 19.8%, Angola at 17.6%, Sudan at 9.6%, Mozambique at 7.9% and Malawi at 7.8%. Many international agencies are gaining increasing interest in emerging modernizing African economies, especially as Africa continues to maintain high economic growth despite current global economic recession.

Regional Variation

While no African nation has joined the ranks of the developed nations in the Organization for Economic Co-operation and Development (OECD) yet, the entire continent is not utterly impoverished and there is considerable variation in its wealth. Arab North Africa has long been closely linked to the economies of Europe and the Middle East. South Africa is by far the continent's wealthiest state, both in GDP per capita and in total GDP, and its neighbours have shared in this wealth. The small but oil-rich states of Gabon and Equatorial Guinea round out the list of the ten wealthiest states in Africa.

The temperate northern and southern ends of the continent are wealthier than tropical sub-Saharan Africa. Within the tropics, East Africa, with its long pre-colonial history of trade and development, has tended to be wealthier and more stable than elsewhere. Islands such as the Seychelles, Réunion, Mauritius, and Cape Verde have remained wealthier than the continental nations, although the unstable Comoros remains poor.

The poorest states are those engaged in or just emerging from civil wars. These include the Democratic Republic of the Congo, Sierra Leone, and Burundi. In recent times, the poorest region has been the Horn of Africa, although it had historically been one of the wealthiest regions of sub-Saharan Africa. Ethiopia in particular had a long and successful history. The current poverty of the region, and the associated famines and wars, have been a problem for decades.

There is considerable internal variation within countries. Urban areas, especially capital cities, are generally wealthier than rural zones. Inequality is pronounced in most African countries; an upper class has a much higher income than the majority of the population.

History

Ancient Egypt was one of the world's most prosperous and advanced civilizations, which began around 3150 BC with the political unification of Upper and Lower Egypt under the first pharaoh, and it developed over the next three millennia. The port of Alexandria, founded by Alexander the Great in 334 BC, was a hub for Mediterranean trade for centuries. Well into the 19th century, Egypt remained one of the most developed regions in the world. Prosperity in the rest of Africa existed in nation states and kingdoms such as the Ghana Empire , Nubia, Ethiopia, and Mali, which had trade routes north to the Mediterranean world and Middle East.

Africans have historically built structures from stone mainly in the Nile Valley in cities like Meroe, Napata, Axum by former Nubian and Ethiopian kingdoms. Most other Sub Saharan African pre-colonial civilizations built mainly out of mud brick, leaving few lasting ruins except Great Zimbabwe. Finding no architectural monuments in most parts of the region, some European explorers and historians long concluded that pre-colonial sub-Saharan Africa was devoid of civilization. It must be noted that racism also blinded some of the European explorers and historians to conclude that pre-colonial Africa was devoid of civilization for example J. Theodore Bent, who researched the origins of Great Zimbabwe stated in *The Ruined Cities of*

Mashonaland (1891) that the ruins revealed either the Phoenicians or the Arabs as builders. Other European researcher favoured a legend that the structures were built to replicate the palace of the Queen of Sheba in Jerusalem. Other theories as to their origin abounded among white settlers and academics, with one racist element in common: they were probably not made by Africans.

New technologies and increasing scales of production made trading easier. For most of the first millennium AD, the Axumite Kingdom had a prosperous trade empire on the eastern horn, where the modern states of Ethiopia and Eritrea lie. Axum had a powerful navy and traded as far as the Byzantine Empire, India, and possibly China. The introduction of the camel by North African Arab conquerors in the 10th century opened trade across the Sahara for the first time. The profits from the gold and salt trades created powerful empires in the western Sahel including the Kingdom of Ghana and the Mali and Kanem-Bornu Empires, where travellers reported vast wealth. Arabs helped build a maritime trade along Africa's east coast, which prospered as Swahili traders exported ivory and slaves across the Indian Ocean.

Further south empires were less common, with the notable exception of Great Zimbabwe. In the Great Lakes region, states such as Rwanda, Burundi, and Buganda became strongly centralized, due to its high population and agricultural surplus.

In the 15th century, Portuguese traders circumvented the Saharan trade route and began to trade directly with Guinea. Other European traders followed, rapidly boosting prosperity in Western Africa. States flourished, including the Kingdom of Benin, Dahomey, and the Ashanti Confederacy. Loose federations of city states such as those of the Yoruba and Hausa were common. However, this wealth was principally based on the slave trade, which collapsed following the abolition of slavery and later European colonization.

Although Europeans were ostensibly committed to developing their colonies, colonial rulers employed a *laissez-faire* strategy during the first decades. It was hoped that European companies would prosper if given a secure operating environment. This only occurred in a few areas with rich resources; the colonial economies hardly grew from the 1890s through the 1920s. The colonies had to pay their own way, receiving little or no development money from Europe. Only in the 1930s, with the rise of Keynesian economics, did the colonial administrations seriously encourage development. However, new projects could not transpire until after the Great Depression and the Second World War.

African economies boomed during the 1950s as growth and international trade multiplied beyond their pre-war levels. The insatiable demand for raw materials in the rebuilding economies of Asia and Europe and the strong growth in North America inflated the price of raw materials. By the end of the colonial era in the 1960s, there was great hope for African self-sufficience and prosperity. However, sporadic growth continued as the newly independent nations borrowed heavily from abroad.

The world economic decline of the 1970s, rising oil prices, corruption, and political instability hit Africa hard. In subsequent decades Africa has steadily become poorer compared to the rest of the world; South America experienced solid growth, and East Asia spectacular growth, during that same period. According to the World Economic Forum, ten percent of the world's poor were African in 1970; by 2000, that figure had risen to 50 percent. Between 1974 and 2000 the average income declined by $200. Beginning in 1976, the Lome Convention and Cotonou Agreement between the European Union and ACP countries, including Sub-Saharan Africa, have structured economic relations between the two regions.

Sectors

It is through the sectors of Africa's work industry that the economy can be maintained. Most of this is contributed to exporting of goods. This is due to the smaller amount of secondary industries available in the continent. In 2009, 87% of Africa's economy was from exportation alone.

Agriculture

Around 60 percent of African workers are employed by the agricultural sector, with about three-fifths of African farmers being subsistence farmers. Subsistence farms provide a source of food and a relatively small income for the family, but generally fail to produce enough to make re-investment possible. Larger farms tend to grow cash crops such as coffee, cotton, cocoa, and rubber. These farms, normally operated by large corporations, cover tens of square kilometres and employ large numbers of labourers.

The situation whereby African nations export crops to the West while millions on the continent starve has been blamed on developed countries including Japan, the European Union and the United States.

These countries protect their own agricultural sectors with high import tariffs and offer subsidies to their farmers, which many contend leads the overproduction of such commodities as grain, cotton and milk. The result of this is that the global price of such products is continually reduced until Africans are unable to compete, except for cash crops that do not grow easily in a northern climate.

Because of these market forces, in Africa excess capacity is devoted to growing crops for export. Thus, when civil unrest or a bad harvest occurs, there is often very little food saved and many starve. Ironically, excess foodstuffs grown in developed nations are regularly destroyed, as it is not economically viable to transport it across the oceans to a market poor in capital. Although cash crops can expand a nation's wealth, there is often a risk that focusing on them rather than staples will lead to food shortages and hunger.

In modern years countries such as Brazil, which has experienced great progress in agricultural production, have agreed to share technology with Africa to greatly increase agricultural production in Africa to make it a more viable trade partner.

Increased investment in African agricultural technology in general has the potential to greatly decrease poverty in Africa. The demand market for African cocoa is currently experiencing an enjoyable price boom. The South African and Ugandan governments have targeted policies to take advantage of the increased demand for certain agricultural products and plan to stimulate agricultural sectors. The African Union has plans to heavily invest in African agriculture and the situation is closely monitored by the UN.

Mining and Drilling

Africa's most valuable exports are minerals and petroleum. A few countries possess and export the vast majority of these resources. The southern nations have large reserves of gold, diamonds, and copper. Petroleum is concentrated in Nigeria, its neighbours, and Libya.

While mining and drilling produce most of Africa's revenues each year, these industries only employ about two million people, a tiny fraction of the continent's population. Profits normally go either to large corporations or to the governments. Both have been known to squander this money on luxuries for the elite or on mega-projects that return little value.

In some cases, these resources have turned out to be detrimental to economic development. Although Congo is rich in minerals, the country remains one of the poorest countries in the world. This is historically due to ownership fights over these minerals, tracing back to the early 1900s. After Congo's independence from Belgium, the colonial government hesitated to leave behind these resources. Congo solicited UN help against Belgium, but that turned out to be a bad idea. In an attempt to get out of the quagmire, Congo sought Soviet assistance. This led the country into deeper trouble, as the country separated into two and a long proxy war between the West and East began. However, countries such as Angola and Uganda are experiencing booms in drilling and oil drilling and manufacture.

Manufacturing

Africa is the least industrialized continent; only South Africa, Egypt, Morocco and Tunisia in general have substantial manufacturing sectors. Despite readily available cheap labour, nearly all of the continent's natural resources are exported for secondary refining and manufacturing. According to the AFDB, about 15% of workers are employed in the industrial sector.

The multinational corporations that control most of the world's major industries and their financiers require political stability before erecting an expensive factory and risk losing that investment through nationalization. An educated populace, good infrastructure and a stable source of electricity are essential to investments. These factors are rare in most countries in

Africa. Other developing regions of the world such as India and China have been more attractive to companies looking to build a new factory or invest in a local enterprise.

Many African states used to limit foreign investment to ensure local majority ownership. Close governmental control over industry further discouraged international investment. Attempts to foster local industry have been hampered by insufficient technology, training, and investment money. The paucity of local markets and the difficulty of transporting goods from major African centres to world markets contribute to the lack of manufacturing outside of South Africa and Egypt.

Both the African Union and the United Nations have outline plans in modern years on how Africa can help itself industrialize and develop significant manufacturing sectors to levels proportional to the African economy in the 1960s with 21st century technology.

This focus on growth and diversification of manufacturing and industrial production, as well as diversification of agricultural production, has fuelled hopes that 21st century will prove to be a century of economic and technological growth for Africa.

This hope coupled with the rise of new leaders in Africa in the future inspired the term "the African Century" referring to the 21st century potentially being the century when Africa's vast untapped labour, capital and resource potentials might become a world player.

This hope in manufacturing and industry is helped by the boom in communications technology and local mining industry in much of sub-Saharan Africa. Namibia has attracted industrial investments in recent years and South Africa has begun offering tax incentives to attract foreign direct investment projects in manufacturing.

Countries such as Mauritius have plans for developing new "green technology" for manufacturing. Developments such as this have huge potential to open new markets for African countries as the demand for alternative "green" and clean technology is predicted to soar in the future as global oil reserves dry up and fossil fuel-based technology becomes more economically nonviable.

Investment and Banking

Banking in Africa has long been problematic. Because local banks are often unstable and corrupt, governments and industry rely on international banks. South Africa and Egypt alone have a thriving banking sector, aided by the international sanctions of the apartheid era, which forced out the once-dominant British banks for the former. In the years after independence, African governments heavily regulated the banking sector and placed strict limits on international competition. In recent decades, banking reform has been a priority of the IMF and World Bank. One important reform was obtaining permission for increased penetration by foreign banks. South Africa and Egypt have been the most successful in attracting local operation of foreign banks. In 2007, Egypt surpassed South Africa as the biggest recipient of FDI recording $11.1 bn. This trend continued in 2008, where Egypt attracted $13.2 bn in FDI.

Encouraging foreign investment in Africa has been difficult. Even Africans are reluctant to invest locally; about forty percent of sub-Saharan African savings are invested in other markets. The IMF and World Bank only lend money after imposing stringent and controversial conditions such as austerity policies. However, China and India have showed exponentially increasing interest in emerging African economies in the 21st century. Investment in Africa by China and African trade with China has increased dramatically in recent years, even regardless of the current world financial crisis. The increased investment in Africa by China has attracted the attention of the European Union and has provoked talks of competitive investment by the EU. Members of the African diaspora abroad, especially in the EU and the United States, have increased efforts to use their businesses to invest in Africa and encourage African investment abroad in the European economy. Remittances from the African diaspora and rising interest in investment from the West will especially be helpful for Africa's least developed and most devastated economies, such as Burundi, Togo and Comoros.

Angola has announced interests in investing in the EU, Portugal in particular. South Africa has attracted increasing attention from the United States as a new frontier of investment in manufacture, financial markets and small business, as has Liberia in recent years with new leadership.

There are two African currency unions: the West African Banque Central des Etats de l'Afrique de l'Ouest (BCEAO) and the Central African Banque des Etats de l'Afrique Central (BEAC). Both use the CFA franc as their legal tender.

Communication and Information Technology

The continent has the largest growth rate of cellular subscribers in the world. African markets are expanding nearly twice as fast as Asian markets. The African cell phone has created a base for cellular banking. Namibia has attracted international attention with new phone services and liberalization of regulatory controls in Kenya are producing a boom in Internet services demand and demand for modern communication technology.

Determinants

The seemingly intractable nature of Africa's poverty runs counter to modern economic theory , leading to debate concerning its root causes. Endemic warfare and unrest, widespread corruption, and despotic regimes are both causes and effects of the continued economic problems.

Geography

Variations of the controversial theory of geographic determinism date back to Montesquieu, though these theories have been simplistic and unscientific until they have recently been revived and refined by academics such as William Masters and Jeffrey Sachs and popular writers such as Jared Diamond.

Africa has the largest hot desert and the second largest tropical forest in the world that hinder building transcontinental means of transportations, hampering its economy. In the centre of the continent, on the western side, an almost impenetrable rainforest impedes the transit of people and goods. The Sahara creates an obvious barrier to trade from the north. Although Africa has great river systems such as the Nile, Niger, Congo, and Zambezi, they do not link the continent into trade routes effectively as happens in Europe and China. Rapids and cataracts block African rivers, requiring development projects to allow navigation. The wet terrain of the interior complicates transport. Few roads are paved and during the wet season unpaved tracks become impassable mud.

Countries in Africa are cut off from the sea more than those on other continents. Africa has more landlocked nations than any other continent, which support a high population density compared to the steppes or plains of North America and Asia. The ridge running from Zimbabwe to Ethiopia has superb volcanic soils and the higher altitude produces a more temperate climate. These enable more interior settlement, but the lack of access to the sea makes international trade harder.

The majority of the world's population and wealth is found in the temperate zone. Historically the vast expanse of Eurasia, almost entirely in the temperate zone (except for the vast tracts that are dry and hot such as the Arabian Peninsula; cold tundra such as in North Asia, and tropical such as subcontinental India, Bangladesh, Thailand, Laos, Bhutan, Burma, Cambodia, Vietnam, Malaysia, and Singapore) was linked by land routes, allowing technologies and ideas to spread from one area over time, aiding innovation.

The agricultural techniques and medicines designed to work in the northern climes may fail in the tropics. This theory could partly explain why temperate South Africa is by far the wealthiest part of Africa, and why other tropical areas in South America and Indonesia share Africa's poverty.

Exceptions exist, Singapore and Brunei among them, however these are isolated spots of wealth. There are no tropical countries in the OECD, apart from Mexico and Australia which have significant tropical sections, and only a handful have a GDP per capita above the world average, again apart from Mexico, Australia, Singapore, Brunei, Malaysia and Thailand. Globally there is a correlation between wealth and climate although it does not fully explain all instances of poverty in Africa.

Africa is well-endowed with natural resources, including gold, diamonds, and oil reserves, but due to several factors including poor governance and global trade policies which place tariffs on finished goods from Africa, few African countries have materially benefited from their mineral wealth. Africa is as well suited to agriculture as any other continent; the volcanic soils of the Great Lakes region are—by some measures—the best in the world.

Disease

Closely linked to geography is the problem of disease in Africa. The tropics are more hospitable to disease than the colder climates. The most significant illness has long been malaria. Since the tropical regions are poorer, pharmaceutical companies are reluctant to invest in curing the diseases of the region. Disease reduces productivity and creates a health care cost burden.

A new problem of vast magnitude is the rise of HIV/AIDS in Sub-Saharan Africa. AIDS, whose spread correlates with poverty, has nevertheless hit hardest in some of the wealthiest African countries, including Botswana, Swaziland, and South Africa. AIDS has decimated or will decimate the working-age population of many states. As parents die or become unable to work, their children must find care elsewhere, adding to the burden of already struggling families and states.

The costs of HIV/AIDS, including importation of anti-retroviral AIDS drugs from the West, is a new burden on many African states, leading to the challenging of drug prices and the manufacture of cheap generic alternatives.

Animal diseases can also harm agriculture and transportation, as most forms of livestock cannot survive the diseases of the region. Historically, sub-Saharan Africans could not use pack animals for trade or work horses for labour, limiting the continent's development.

Colonialism

The economic impact of the colonization of Africa has been debated. Africa acquired its greatest relative wealth in the 1960s, just prior to decolonization. African countries have yet to return to those levels of wealth. Some see this as evidence that colonialism helped local economies by creating a cyclical economic link with the ruling colonial power, and with independence this link was broken.

To achieve wealth during the colonial period, imperial overseers geared the economies of Africa towards exporting raw materials. Egypt produced cotton, Ruanda-Urundi was almost completely dedicated to growing coffee, and Upper Volta specialized in palm oil. Basing an entire nation's wealth on one commodity in this way would have debilitating effects later. These mono cultures left national economies extremely vulnerable to price swings, making economic planning difficult. Some writers, such as Walter Rodney in his influential book *How Europe Underdeveloped Africa*, argue that these colonial policies are directly responsible for many of Africa's modern problems.

Other post-colonial scholars, most notably Frantz Fanon, have argued that the true effects of colonialism are psychological and that domination by a foreign power creates a lasting sense of inferiority and subjugation that creates a barrier to growth and innovation.

Once independent, African states saw an exodus of European administrators and consequently lacked individuals with the training or education to operate the government they had inherited. For instance, the massive area of French Equatorial Africa was divided into four independent nations, but was home to only five locals who were university graduates.

One method of gauging the effects of colonialism on the economies of Africa is to compare the results of different colonial policies implemented by the European powers. Regions where the economy was plundered, such as the Raubwirtschaft policies of Leopold II in the Congo Free State, have not prospered. The long reluctance of Portugal to surrender its colonies, leading to long wars of independence, had an obvious negative effect on Mozambique and Angola.

The countries formerly under French control have fared much better, and those under British dominion were the most successful. This inequality may be due to other factors than economic policy. Britain, at the time of the Scramble for Africa, was the world's greatest power and could thus cherry-pick the wealthiest parts of the continent. The French, with their mighty navy, could also occupy prosperous areas, while the Belgians were forced to take the economically disadvantaged interior.

Africa as a whole has not prospered compared with other colonised regions in Asia and the Americas. At the end of the Second World War the Americas were economically the strongest of the colonised regions; in the span of one generation, former colonies in Asia have become economic powerhouses.

Borders

In the Scramble for Africa, national boundaries in sub-Saharan Africa were established by Europeans using latitude and longitude rather than natural borders. This separated population centres from their supplies of food and natural resources. The artificial borders of modern African states cut across cultural, tribal, linguistic and religious boundaries, creating ethnic and religious cleavages which impede national unity and induce internal violence.

However, those states that preserved pre-colonial boundaries have been no more successful. Few countries in Africa have more troubled recent histories than Rwanda and Burundi, although their borders are almost identical to those of the prosperous kingdoms from which they are descended. The ancient and only briefly occupied state of Ethiopia is one of the poorest on the continent. Elsewhere, the Americas were also divided up by Europeans along arbitrary borders and yet those continents remain economically far more successful than Africa.

Africa is a much divided continent with many small countries. Successful economic growth requires regional cooperation, which political tensions make difficult. To be effective, foreign aid must be multilateral, making it harder to base aid upon the performance of local governments.

Language Diversity

African countries suffer from communication difficulties caused by language diversity. Greenberg's diversity index is the chance that two randomly selected people would have different mother tongues. Out of the most diverse 25 countries according to this index, 18 (72%) are African. This includes 12 countries for which Greenberg's diversity index exceeds 0.9, meaning that a pair of randomly selected people will have less than 10% chance of having the same mother tongue. The primary language of government, political debate, academic discourse, and administration is often the language of the former colonial powers—English, French, or Portuguese. Only an elite minority speak these European languages fluently enough to participate in these institutions without intermediaries, further disenfranchising the majority population.

Governance

The political situation in Africa perpetuates the intractable nature of African poverty. Democracy in Africa has not been historically successful, almost always supplanted by centralized authoritarian rule such as military dictatorships. Although some rulers worked to improve the lot of their nation's citizens, others used power purely for their own benefit. Among the most notorious was Mobuto Sese Seko of Zaire, whose regime has been called a kleptocracy due to its looting of the nation's wealth. According to international measures, the economies of Africa generally rank among the most corrupt. Bribery and graft abound, due to poverty and poorly handled de-colonization, and the superpowers' (Soviet Union and United States) practice during the Cold War of supporting any ruler with the desired political alignment, regardless of their managerial practices or human rights records.

Dependency theory asserts that the wealth and prosperity of the superpowers and their allies in Europe, North America and East Asia is dependent upon the poverty of the rest of the world, including Africa. Economists who subscribe to this theory believe that poorer regions must break their trading ties with the developed world in order to prosper.

Less radical theories suggest that economic protectionism in developed countries hampers Africa's growth. When developing countries have harvested agricultural produce at low cost, they generally do not export as much as would be expected. Abundant farm subsidies and high import tariffs in the developed world, most notably those set by Japan, the European Union's Common Agricultural Policy, and the United States Department of Agriculture, are thought to be the cause. Although these subsidies and tariffs have been gradually reduced, they remain high.

Local conditions also affect exports. In addition to lack of mechanization or other infrastructure, state over-regulation in several African nations can prevent their own exports from becoming competitive. Research in Public Choice economics such as that of Jane Shaw suggest that protectionism operates in tandem with heavy State intervention combining to depress economic development. Farmers subject to import and export restrictions cater to localized markets, exposing them to higher market volatility and fewer opportunities. When subject to uncertain market conditions farmers press for governmental intervention to suppress competition in their markets, resulting in competition being driven out of the market. As competition is driven out of the market farmers innovate less and grow less food further undermining economic performance.

Although Africa and Asia had similar levels of income in the 1960s, Asia has since outpaced Africa. One school of economists argues that Asia's superior economic development lies in local investment. Corruption in Africa consists primarily of extracting economic rent and moving the resulting financial capital overseas instead of investing at home; the stereotype of African dictators with Swiss bank accounts is often accurate. Asian dictators such as Suharto often take a cut on everything, necessitating bribery, but enable development through infrastructure investment and the social stability created by law and order. University of Massachusetts researchers estimate that from 1970 to 1996, capital flight from 30 sub-Saharan countries totalled $187bn, exceeding those nations' external debts. This disparity in development is consistent with the model theorized by economist Mancur Olson. Because governments were politically unstable and new governments often confiscated their predecessors' assets, officials would stash their wealth abroad, out of reach of any future expropriation.

Corruption encouraged social inequality, because the wealthy elite not only avoided investing at home, but also imported most of its consumption. Desirable luxury goods were generally not locally available. This hindered the development of national markets. Historically, economic development is closely linked to the creation of a middle class with enough income to save and invest but limited influence on governance. In countries without such a middle class, development is all but impossible, beyond resource extraction..

War

Since independence, Africa has seen dozens of wars, both civil and international. This has contributed to poverty because states have spent their scarce resources on military equipment and supplies. Development has suffered, since warfare has scared off foreign investors, destroyed infrastructure, and created lasting animosities.

Much conflict was enabled by the Cold War. The countries of the Western and Eastern blocs leveraged foreign aid money to coax countries into their camp. Aid was tied to the purchase of military weapons, and donor countries ignored misappropriation of the funds. Corruption became endemic, hampering economic development. Proxy conflicts erupted in Africa when each bloc would fund and assisted rebel or sectarian groups under the control of the opposing bloc.

Violence in Africa has increased following the Cold War, despite the slashing of foreign aid spending in developed countries. Civil wars have raged throughout the African Great Lakes region, Somalia, Sudan, Mozambique, Liberia, Sierra Leone, Ivory Coast, and Guinea-Bissau. International wars include the First and Second Congo Wars between the Democratic Republic of the Congo and its neighbours, and conflict between Ethiopia and its former province Eritrea.

Effects of Widespread Poverty

Africa's economic malaise is self-perpetuating, as it engenders more of the disease, warfare, misgovernment, and corruption that created it in the first place. Other effects of poverty have similar consequences. The most direct consequence of low GDP is Africa's low standard of living and quality of life. Except for a wealthy elite and the more prosperous peoples of South Africa and the Maghreb, Africans have very few consumer goods. Quality of life does not correlate exactly with a nation's wealth. Angola, for instance, reaps large sums annually from its diamond mines, but after years of civil war, conditions there remain poor. Radios, televisions, and automobiles are rare luxuries. Most Africans are on the far side of the Digital Divide and are cut off from communications technology and the Internet. Quality of life and human development are also low. African nations dominate the lower reaches of the UN Human Development Index. Infant mortality is high, while life expectancy, literacy, and education are all low. The UN also lowers the ranking of African states because the continent sees greater inequality than any other region. The best educated often choose to leave the continent for the West or the Persian

Gulf to seek a better life. Catastrophes cause deadly periods of great shortages. The most damaging are the famines that have regularly hit the continent, especially the Horn of Africa. These have been caused by disruptions due to warfare, years of drought, and plagues of locusts.

An average African faced annual inflation of over 60% from 1990 until 2002 in those few countries that account for inflation. At the high end, Angola and the Democratic Republic of the Congo both saw triple-digit inflation throughout the period. Most African states saw inflation of around 10% per year.

There are no reliable numbers for unemployment in most African nations, but it is an important problem. Major cities like Lagos, Kinshasa, and Nairobi notably have large slums of the unemployed and underemployed.

Environmental degradation occurs on many fronts. Farmers on the verge of starvation are unlikely to be concerned about the fate of the rainforest in their pursuit of new land, and starving people do not often consider the rarity of an animal before eating it. Along the length of the Sahel, deforestation and overgrazing has caused increased desertification as the Sahara spreads south. Profits from the sale in the West of rare animals, ivory from elephants, and timber encourage illegal poaching. Local governments have little money to devote to protecting the environment.

Attempts at Promoting Growth

The relative economic failure of Africa has long been an important issue both in Africa and abroad. Many attempts at solving Africa's poverty have been attempted, but few have had any great degree of success.

Socialism

In the years immediately after independence many nations saw the rapid industrialisations of the Soviet Union and China under communism as models to follow. This led to command economies and major investment in heavy industries such as coal and steel production to stimulate growth, but this approach had little success. Only a handful of states formally adopted socialism and even fewer turned to outright Marxism. Everywhere government intervention in the economy was seen as necessary for growth, especially since private companies and investors were unlikely to invest in the region.

Often the approach of governments in Africa was to borrow heavily from abroad and use this aid to grow the economy to a level that the loans could be paid off. Sporadic growth during the years after independence continued. The countries focused on exports to pay for these development efforts. The 1973 energy crisis hit sub-Saharan Africa as hard as anywhere in the world. While some nations were net exporters, most were heavily reliant on imported fuels. Economies quickly began to falter and events such as famines hit Africa in the 1980s. The collapse of the Soviet Union, which had supported socialist and collectivist projects throughout the continent, undermined the legitimacy of such an approach, while it also meant that there were no longer any sources of international aid to help pursue this approach. However, there are a few successful socialist endeavours in Africa which have led to growth or increasing wealth in a small handful of African countries such as Libya and Angola.

Liberalism

Thus in the 1980s, socialist ideas were discarded throughout almost the entire continent as capitalism became seen as the route to salvation in what became known as the Washington Consensus. By 1990, forty of the nations of Sub-Saharan Africa had agreed to follow rigorous IMF restructuring plans. IMF recommendations saw the continent's currencies drop by an average of 50%, the selling off of government-owned industries, and the slashing of government spending. After twenty years, however, these methods have seen as little success as the socialist approaches of the previous era. Average growth increased from 2.3% per annum to 2.8%. Only a handful of African states reached new levels of wealth, and many others became poorer over the course of the 1990s. Today there is a great deal of controversy on why this failed. One school of thought is that the reforms failed because they were only economic in nature and without democracy and the rule of law development cannot occur.

Yet another school of thought attributes some of Africa's problems to insufficient liberalization. It has been pointed out that while the developed world has insisted that Africa open its markets and eliminate public subsidies, this has been one-sided as the developed world has not opened its markets to agricultural goods from Africa nor has it eliminated agricultural subsidies . At the GATT free trade talks, the African leaders repeatedly request that the developed nations abolish the subsidies they provide their farmers and open their markets to African agricultural goods. It has been argued that the abolition of the subsidy would have three beneficial effects for the developing world and Africa:

- The developed nations would produce less food locally, therefore providing a larger export market for developing countries.
- Food prices would rise without the artificial subsidy and therefore would increase profits for food exports from the developing world.
- The developing nations could adopt a more balanced agriculture policy, producing food and grain for export; this would provide a surplus that would shield countries from famine.

Autarky

The pursuit of self-sufficiency as advocated by dependency theory has been given limited trials in several African countries. In the 1980s, Nigeria banned the importation of many foodstuffs to stimulate domestic production. The Lagos Plan of Action of 1982 called for Africa as a whole to block imports from the rest of the world, but few countries followed through on the idea. Eventually even Nigeria agreed to limited liberalization.

Foreign Aid

Since independence there has been a constant flow of foreign aid into Africa. The benefits of this aid have been mixed. In many cases much of this aid was misappropriated by unscrupulous leaders. During the Cold War the main goal of much of the aid money was to win the allegiance of these rulers, and so their misappropriation of the aid was at the very least overlooked. Since the end of the Cold War almost all developed countries have slashed foreign aid spending. Many also allege that the aid that was not stolen was long misdirected. For many decades the leading notion of development was government supervised mega-projects; today many believe that small grants to local businesses would be more effective. One example of foreign aid which has come under considerable criticism is food aid. In some circles, it is believed that food aid does not solve any fundamental problems and can also lead to a dependency on outside assistance, as well as hindering the development of indigenous industries. Food shipments in case of dire local shortage are generally uncontroversial; but as Amartya Sen has shown, most famines involve a local lack of income rather than of food. In such situations, food aid - as opposed to financial aid - has the effect of destroying local agriculture and serves mainly to benefit Western agribusiness which are vastly overproducing food as a result of agricultural subsidies. Historically, food aid is more highly correlated with excess supply in Western countries than with the needs of developing countries.

Debt Relief

Advocacy for debt relief has become widespread. Each year Africa sends more money to Western bankers in interest on its debts than it receives in foreign aid from these countries. Debt relief is not a panacea, but relieving some of the burden, especially of debts that were run up by regimes for their own benefit, may help the economies of Africa grow and prosper.

Arguments against full and unconditional debt relief include:

- Debt relief punishes nations which have managed borrowing well and do not need debt relief.
- Unconditional debt relief will not necessarily cause nations to spend more in social programs and services, on the one hand, or to solve their financial problems without stifling the economy with the need for more taxes, on the other hand.
- Debt relief may make it more difficult for nations to receive credit in the future.

It has been suggested that any debt relief policy be conditional upon a commensurate reduction in aid.

The Heavily Indebted Poor Countries initiative was launched in 1996; in the 2000s, eligible countries began receiving debt relief. Future foreign aid is planned to be more heavily weighted toward grants than loans.

International Unions

The African Union is the largest international economic grouping on the continent. The confederation's goals include the creation of a free trade area, a customs union, a single market, a central bank, and a common currency, thereby establishing economic and monetary union. The current plan is to establish an African Economic Community with a single currency by 2023. The African Investment Bank is meant to stimulate development. The AU plans also include a transitional African Monetary Fund leading to an African Central Bank. Some parties support development of an even more unified United States of Africa.

International monetary and banking unions include:

- Central Bank of West African States
- Bank of Central African States
- Common Monetary Area.

Although it has abundant natural resources, Africa remains the world's poorest and most underdeveloped continent, due to a variety of causes that may include the spread of deadly diseases and viruses (notably HIV/AIDS and malaria), corrupt governments that have often committed serious human rights violations, failed central planning, high levels of illiteracy, lack of access to foreign capital, and frequent tribal and military conflict (ranging from guerrilla warfare to genocide). According to the United Nations' Human Development Report in 2003, the bottom 25 ranked nations (151st to 175th) were all African.

Poverty, illiteracy, malnutrition and inadequate water supply and sanitation, as well as poor health, affect a large proportion of the people who reside in the African continent. In August 2008, the World Bank announced revised global poverty estimates based on a new international poverty line of $1.25 per day (versus the previous measure of $1.00). 80.5% of the Sub-Saharan Africa population was living on less than $2.50 (PPP) a day in 2005, compared with 85.7% for India.

The new figures confirm that sub-Saharan Africa has been the least successful region of the world in reducing poverty

($1.25 per day); some 50% of the population living in poverty in 1981 (200 million people), a figure that rose to 58% in 1996 before dropping to 50% in 2005 (380 million people). The average poor person in sub-Saharan Africa is estimated to live on only 70 cents per day, and was poorer in 2003 than he or she was in 1973 indicating increasing poverty in some areas. Some of it is attributed to unsuccessful economic liberalization programs spearheaded by foreign companies and governments, but other studies and reports have cited bad domestic government policies more than external factors.

From 1995 to 2005, Africa's rate of economic growth increased, averaging 5% in 2005. Some countries experienced still higher growth rates, notably Angola, Sudan and Equatorial Guinea, all three of which had recently begun extracting their petroleum reserves or had expanded their oil extraction capacity. The continent has 90% of the world's cobalt, 90% of its platinum, 50% of its gold, 98% of its chromium, 70% of its tantalite, 64% of its manganese and one-third of its uranium. The Democratic Republic of the Congo (DRC) has 70% of the world's coltan and most mobile phones in the world have coltan in them. The DRC also has more than 30% of the world's diamond reserves. Guinea is the world's largest exporter of bauxite. As the growth in Africa has been driven mainly by services and not manufacturing or agriculture, it has been growth without jobs and without reduction in poverty levels. In fact, the food security crisis of 2008 which took place on the heels of the global financial crisis has pushed back 100 million people into food insecurity.

In recent years, the People's Republic of China has built increasingly stronger ties with African nations. In 2007, Chinese companies invested a total of US$1 billion in Africa.

Demographics

Africa's population has rapidly increased over the last 40 years, and consequently it is relatively young. In some African states half or more of the population is under 25 years of age. African population grew from 221 million in 1950 to 1 billion in 2009.

Speakers of Bantu languages (part of the Niger-Congo family) are the majority in southern, central and East Africa proper. But there are also several Nilotic groups in East Africa, and a few remaining indigenous Khoisan ('San' or 'Bushmen') and Pygmy peoples in southern and central Africa, respectively. Bantu-speaking Africans also predominate in Gabon and Equatorial Guinea, and are found in parts of southern Cameroon. In the Kalahari Desert of Southern Africa, the distinct people known as the Bushmen (also "San", closely related to, but distinct from "Hottentots") have long been present. The San are physically distinct from other Africans and are the indigenous people of southern Africa. Pygmies are the pre-Bantu indigenous peoples of central Africa.

The peoples of North Africa comprise two main groups; Berber and Arabic-speaking peoples in the west, and Egyptians in the east. The Arabs who arrived in the seventh century introduced the Arabic language and Islam to North Africa. The Semitic Phoenicians, the Iranian Alans, the European Greeks, Romans and Vandals settled in North Africa as well. Berbers still make up the majority in Morocco, while they are a significant minority within Algeria. They are also present in Tunisia and Libya. The Tuareg and other often-nomadic peoples are the principal inhabitants of the Saharan interior of North Africa. Nubians are a Nilo-Saharan-speaking group (though many also speak Arabic), who developed an ancient civilisation in northeast Africa.

Some Ethiopian and Eritrean groups (like the Amhara and Tigrayans, collectively known as "Habesha") speak Semitic languages. The Oromo and Somali peoples speak Cushitic languages, but some Somali clans trace their founding to legendary Arab founders. Sudan and Mauritania are divided between a mostly Arabized north and a native African south (although the "Arabs" of Sudan clearly have a predominantly native African ancestry themselves). Some areas of East Africa, particularly the island of Zanzibar and the Kenyan island of Lamu, received Arab Muslim and Southwest Asian settlers and merchants throughout the Middle Ages and in antiquity.

Prior to the decolonisation movements of the post-World War II era, Whites were represented in every part of Africa. Decolonisation during the 1960s and 1970s often resulted in the mass emigration of European-descended settlers out of Africa – especially from Algeria (*pieds-noirs*), Morocco, Kenya, Congo, Angola, Mozambique and Rhodesia. Nevertheless, White

Africans remain an important minority in many African states. The African country with the largest White African population is South Africa. The Afrikaners, the Anglo-Africans and the Coloureds are the largest European-descended groups in Africa today.

European colonisation also brought sizeable groups of Asians, particularly people from the Indian subcontinent, to British colonies. Large Indian communities are found in South Africa, and smaller ones are present in Kenya, Tanzania, and some other southern and East African countries. The large Indian community in Uganda was expelled by the dictator Idi Amin in 1972, though many have since returned. The islands in the Indian Ocean are also populated primarily by people of Asian origin, often mixed with Africans and Europeans. The Malagasy people of Madagascar are an Austronesian people, but those along the coast are generally mixed with Bantu, Arab, Indian and European origins. Malay and Indian ancestries are also important components in the group of people known in South Africa as Cape Coloureds (people with origins in two or more races and continents). During the 20th century, small but economically important communities of Lebanese and Chinese have also developed in the larger coastal cities of West and East Africa, respectively.

Languages

By most estimates, well over a thousand languages (UNESCO has estimated around two thousand) are spoken in Africa. Most are of African origin, though some are of European or Asian origin. Africa is the most multilingual continent in the world, and it is not rare for individuals to fluently speak not only multiple African languages, but one or more European ones as well. There are four major language families indigenous to Africa.

- The *Afro-Asiatic* languages are a language family of about 240 languages and 285 million people widespread throughout the Horn of Africa, North Africa, the Sahel, and Southwest Asia.
- The *Nilo-Saharan* language family consists of more than a hundred languages spoken by 30 million people. Nilo-Saharan languages are spoken by Nilotic tribes in

Chad, Ethiopia, Kenya, Sudan, Uganda, and northern Tanzania.

- The *Niger-Congo* language family covers much of Sub-Saharan Africa and is probably the largest language family in the world in terms of different languages.
- The *Khoisan* languages number about fifty and are spoken in Southern Africa by approximately 120,000 people. Many of the Khoisan languages are endangered. The Khoi and San peoples are considered the original inhabitants of this part of Africa.

Following the end of colonialism, nearly all African countries adopted official languages that originated outside the continent, although several countries also granted legal recognition to indigenous languages (such as Swahili, Yoruba, Igbo and Hausa). In numerous countries, English and French are used for communication in the public sphere such as government, commerce, education and the media. Arabic, Portuguese, Afrikaans and Malagasy are examples of languages that trace their origin to outside of Africa, and that are used by millions of Africans today, both in the public and private spheres.

Culture

Some aspects of traditional African cultures have become less practiced in recent years as a result of years of neglect and suppression by colonial and post-colonial regimes. There is now a resurgence in the attempts to rediscover and revalourise African traditional cultures, under such movements as the African Renaissance, led by Thabo Mbeki, Afrocentrism, led by a group of scholars, including Molefi Asante, as well as the increasing recognition of traditional spiritualism through decriminalization of Vodou and other forms of spirituality. In recent years, traditional African culture has become synonymous with rural poverty and subsistence farming.

Visual Art and Architecture

African art and architecture reflect the diversity of African cultures. The oldest existing examples of art from Africa are 82,000-year-old beads made from *Nassarius* shells that were found in the Aterian levels at Grotte des Pigeons, Taforalt, Morocco. The Great Pyramid of Giza in Egypt was the world's tallest structure for 4,000 years, until the completion of Lincoln Cathedral around the year 1300. The stone ruins of Great Zimbabwe are also noteworthy for their architecture, and the complexity of monolithic churches at Lalibela, Ethiopia, of which the Church of Saint George is representative.

Music and Dance

Egypt has long been a cultural focus of the Arab world, while remembrance of the rhythms of sub-Saharan Africa, in particular West Africa, was transmitted through the Atlantic slave trade to modern samba, blues, jazz, reggae, hip hop, and rock. The 1950s through the 1970s saw a conglomeration of these various styles with the popularization of Afrobeat and Highlife music.

Modern music of the continent includes the highly complex choral singing of southern Africa and the dance rhythms of the musical genre of soukous, dominated by the music of the Democratic Republic of Congo. Indigenous musical and dance traditions of Africa are maintained by oral traditions, and they are distinct from the music and dance styles of North Africa and Southern Africa. Arab influences are visible in North African music and dance and, in Southern Africa, Western influences are apparent due to colonisation.

Sports

Fifty-three African countries have football (soccer) teams in the Confederation of African Football, while Cameroon, Nigeria, Senegal, and Ghana have advanced to the knockout stage of recent FIFA World Cups. South Africa will host the 2010 World Cup tournament, and will be the first African country to do so.

Cricket is popular in some African nations. South Africa and Zimbabwe have Test status, while Kenya is the leading non-test team in One-Day International cricket and has attained permanent One-Day International status. The three countries jointly hosted the 2003 Cricket World Cup. Namibia is the other African country to have played in a World Cup. Morocco in northern Africa has also hosted the 2002 Morocco Cup, but the national team has never qualified for a major tournament. Rugby is a popular sport in South Africa.

Religion

Africans profess a wide variety of religious beliefs and statistics on religious affiliation are difficult to come by since they are too sensitive a topic for governments with mixed populations. According to the World Book Encyclopedia, Islam is the largest religion in Africa, followed by Christianity. According to Encyclopedia Britannica, 45% of the population are Muslims, 40% are Christians and less than 15% are non-religious or follow African religions. A small number of Africans are Hindu, Bahai, or have beliefs from the Judaic tradition. Examples of African Jews are the Beta Israel, Lemba peoples and the Abayudaya of Eastern Uganda.

Territories and Regions

The countries in this table are categorised according to the scheme for geographic subregions used by the United Nations, and data included are per sources in cross-referenced articles. Where they differ, provisos are clearly indicated.

11

Algeria

Algeria, officially the People's Democratic Republic of Algeria, is a country in North Africa. In terms of land area, it is the largest country on the Mediterranean Sea, the second largest on the African continent after Sudan, and the eleventh-largest country in the world.

Algeria is bordered in the northeast by Tunisia, in the east by Libya, in the west by Morocco, in the southwest by Western Sahara, Mauritania, and Mali, in the southeast by Niger, and in the north by the Mediterranean Sea. Its size is almost 2,400,000 square kilometres (930,000 sq mi), and it has an estimated population of about 35.7 million (2010). The capital of Algeria is Algiers.

Algeria is a member of the Arab League, United Nations, African Union, and OPEC. It is also a founding member of the Arab Maghreb Union.

Etymology

The name of the country is derived from the city of Algiers. In classical times northern Algeria was known as Numidia, which included parts of modern day western Tunisia and eastern Morocco.

History

Ancient History: Berber Algeria

In Antiquity Algeria was known as the Numidia kingdom and its people were called Numidians. The kingdom of Numidia had early relations with Carthaginians, Romans and Ancient

Greeks, the region was considered a fertile area, and Numidians were known for their fine cavalry.

Author Terrence McKenna has hypothesised Algeria as the source of the story of the Garden of Eden and the birth of humanity. Before the warming brought on by the Holocene Climatic Optimum around 7,000 years ago, the region contained vast grasslands which, along with the representations of cattle in the Tassili Plateau art, suggests the existence of early forms of pastoral agriculture. This would seem to be a logical precursor to the crop-based agriculture that developed in the Middle East in the agricultural revolution thousands of years later.

The ancient paintings are also a clear indication of a form of Shamanism or religion based on the use of psychedelic mushrooms. This is another reason why the region was proposed by McKenna as the cradle of culture and civilization, as the visions induced by these mushrooms give a powerful impulse towards art, painting and the sense of contact with the supernatural that is the basis of religious belief.

The indigenous peoples of northern Africa eventually coalesced into a distinct native population, the Berbers.

After 1000 BCE, the Carthaginians began establishing settlements along the coast. The Berbers seized the opportunity offered by the Punic Wars to become independent of Carthage, and Berber kingdoms began to emerge, most notably Numidia.

In 200 BCE, they were once again taken over, this time by the Roman Republic. When the Western Roman Empire collapsed in 476 AD, Berbers became independent again in many regions, while the Vandals took control over other areas, where they remained until expelled by the Byzantine general Belisarius under the direction of Emperor Justinian I. The Byzantine Empire then retained a precarious grip on the east of the country until the coming of the Arabs in the eighth century.

Middle Ages

Berber people controlled much of the Maghreb region throughout the Middle Ages. The Berbers were made up of several tribes. The two main branches were Botr and Barnes, who were themselves divided into tribes, and again into sub-tribes. Each region of the Maghreb contained several tribes (for example, Sanhadja, Houaras, Zenata, Masmouda, Kutama, Awarba, and Berghwata). All these tribes had independence and made territorial decisions.

Several Berber dynasties emerged during the Middle Ages in the Maghreb, Sudan, Andalusia, Italy, Mali, Niger, Senegal, Egypt, and other nearby lands. Ibn Khaldun provides a table summarizing the Berber dynasties: Zirid, Banu Ifran, Maghrawa, Almoravid, Hammadid, Almohad, Merinid, Abdalwadid, Wattasid , Meknassa and Hafsid dynasties.

Arrival of Islam

The Muslim Arab armies arrived in Algeria in the mid-7th century, they conquered Algeria from its former Berber rulers and the Byzantines. After the fall of the Umayyad Arab Dynasty 751, numerous local dynasties emerged. Amongst those dynasties were the Aghlabids, Almohads, Abdalwadid, Zirids, Rustamids, Hammadids, Almoravids, and the Fatimids.

Having converted the Kutama of Kabylie to its cause, the Shia Fatimids overthrew the Rustamids, and conquered Egypt, leaving Algeria and Tunisia to their Zirid vassals. When the latter rebelled, the Shia Fatimids sent in the Banu Hilal, a populous Arab tribe, to weaken them.

Spanish Enclaves

The Spanish expansionist policy in North Africa began with the Catholic monarchs Isabella I of Castile and Ferdinand II of Aragon and their regent Cisneros, once the *Reconquista* of the Iberian Peninsula was completed. Several towns and outposts on the Algerian coast were conquered and occupied by the Spanish Empire: Mers El Kebir (1505), Oran (1509), Algiers (1510) and Bugia (1510). On 15 January 1510 the King of Algiers, Samis El Felipe, was forced into submission to the king of Spain.

King El Felipe called for help from the corsairs Hayreddin Barbarossa and Oruc Reis who previously helped Andalusian Muslims and Jews escape from Spanish oppression in 1492. In 1516, Oruc Reis conquered Algiers with the support of 1,300 Turkish soldiers on board 16 Galliots and became ruler, with Algiers joining the Ottoman Empire.

The Spaniards left Algiers in 1529, Bujia in 1554, Mers El Kebir and Oran in 1708. The Spanish returned in 1732 when the armada of the Duke of Montemar was victorious in the Battle of Ain-el-Turk; Spain recaptured Oran and Mers El Kebir. Both cities were held until 1792, when they were sold by King Charles IV of Spain to the Bey of Algiers.

Ottoman Rule

Algeria was made part of the Ottoman Empire by Hayreddin Barbarossa and his brother Aruj in 1517. After the death of Oruc Reis in 1518, his brother Suneel Basi succeeded him. The Sultan Selim I sent him 6000 soldiers and 2000 janissaries with which he liberated most of the Algerian territory taken by the Spanish, from Annaba to Mostaganem. Further Spanish attacks led by Hugo of Moncada in 1519 were also pushed back. In 1541 Charles V, emperor of the Holy Roman Empire, attacked Algiers with a convoy of 65 warships, 451 ships and 23000 men including 2000 riders, but it was a total failure, and the Algerian leader Hassan Agha became a national hero. Algiers then became a great military power.

The Ottomans established Algeria's modern boundaries in the north and made its coast a base for the Ottoman corsairs; their privateering peaked in Algiers in the 1600s. Piracy on American vessels in the Mediterranean resulted in the First (1801–1805) and Second Barbary Wars (1815) with the United States. The pirates forced the people on the ships they captured into slavery; when the pirates attacked coastal villages in southern and Western Europe the inhabitants were forced into the Arab slave trade.

The Barbary pirates, also sometimes called Ottoman corsairs or the Marine Jihad, were Muslim pirates and privateers that operated from North Africa, from the time of the Crusades until the early 19th century. Based in North African ports such as Tunis in Tunisia, Tripoli in Libya, Algiers in Algeria, Sale and other ports in Morocco, they preyed on Christian and other non-Islamic shipping in the western Mediterranean Sea.

Their stronghold was along the stretch of northern Africa known as the Barbary Coast (a medieval term for the Maghreb after its Berber inhabitants), but their predation was said to extend throughout the Mediterranean, south along West Africa's

Atlantic seaboard, and into the North Atlantic as far north as Iceland and the United States. They often made raids, called *Razzias*, on European coastal towns to capture Christian slaves to sell at slave markets in places such as Turkey, Egypt, Iran, Algeria and Morocco. According to Robert Davis, from the 16th to 19th century, pirates captured 1 million to 1.25 million Europeans as slaves. These slaves were captured mainly from seaside villages in Italy, Spain and Portugal, and from farther places like France, England, Ireland, the Netherlands, Germany, Poland, Russia, Scandinavia and even Iceland, India, Southeast Asia and North America.

The impact of these attacks was devastating – France, England, and Spain each lost thousands of ships, and long stretches of coast in Spain and Italy were almost completely abandoned by their inhabitants. Pirate raids discouraged settlement along the coast until the 19th century.

The most famous corsairs were the Ottoman *Barbarossa* ("Redbeard") brothers — Hayreddin and his older brother Oruc Reis — who took control of Algiers in the early 16th century and turned it into the centre of Mediterranean piracy and privateering for three centuries, as well as establishing the Ottoman Empire's presence in North Africa which lasted four centuries.

Other famous Ottoman privateer-admirals included Turgut Reis (known as Dragut in the West), Kurtoglu (known as Curtogoli in the West), Kemal Reis, Salih Reis, Nemdil Reis and Koca Murat Reis. Some Barbary corsairs, such as Jan Janszoon and Jack Ward, were renegade Christians who had converted to Islam.

In 1544, Hayreddin captured the island of Ischia, taking 4,000 prisoners, and enslaved some 9,000 inhabitants of Lipari, almost the entire population. In 1551, Turgut Reis enslaved the entire population of the Maltese island Gozo, between 5,000 and 6,000, sending them to Libya. In 1554, pirates sacked Vieste in southern Italy and took an estimated 7,000 slaves. In 1555, Turgut Reis sacked Bastia, Corsica, taking 6000 prisoners.

In 1558, Barbary corsairs captured the town of Ciutadella (Minorca), destroyed it, slaughtered the inhabitants and took

3,000 survivors to Istanbul as slaves. In 1563, Turgut Reis landed on the shores of the province of Granada, Spain, and captured coastal settlements in the area, such as Almunecar, along with 4,000 prisoners. Barbary pirates often attacked the Balearic Islands, and in response many coastal watchtowers and fortified churches were erected. The threat was so severe that the island of Formentera became uninhabited.

From 1609 to 1616, England lost 466 merchant ships to Barbary pirates. In the 19th century, Barbary pirates would capture ships and enslave the crew. Latterly American ships were attacked. During this period, the pirates forged affiliations with Caribbean powers, paying a "license tax" in exchange for safe harbour of their vessels. One American slave reported that the Algerians had enslaved 130 American seamen in the Mediterranean and Atlantic from 1785 to 1793.

The cities of North Africa were especially hard hit by the plague. 30,000–50,000 died in Algiers in 1620–21, 1654–57, 1665, 1691, and 1740–42.

French Rule

On the pretext of a slight to their consul, the French invaded and captured Algiers in 1830. The conquest of Algeria by the French was long and resulted in considerable bloodshed. A combination of violence and disease epidemics caused the indigenous Algerian population to decline by nearly one-third from 1830 to 1872.

Between 1825 and 1847 50,000 French people emigrated to Algeria, but the conquest was slow because of intense resistance from such people as Emir Abdelkader, Cheikh Mokrani, Cheikh Bouamama, the tribe of Ouled Sid Cheikh, whose relationships with the French vacillated from cooperation to resistance, Ahmed Bey and Fatma N'Soumer. Indeed, the conquest was not technically complete until the early 1900s when the last Tuareg were conquered.

Meanwhile, however, the French made Algeria an integral part of France. Tens of thousands of settlers from France, Spain, Italy, and Malta moved in to farm the Algerian coastal plain and occupied significant parts of Algeria's cities.

These settlers benefited from the French government's confiscation of communal land and the application of modern agricultural techniques that increased the amount of arable land. Algeria's social fabric suffered during the occupation: literacy plummeted, while land development uprooted much of the population.

Starting from the end of the 19th century, people of European descent in Algeria (or natives like Spanish people in Oran), as well as the native Algerian Jews (typically Mizrachi and sometimes Sephardic in origin), became full French citizens. Formally Algeria as a French territory was member of European Union from the founding of the European Community of Coal and Steel (ECSC) in 1952. Formal membership ended with independence in 1962.

After Algeria's 1962 independence, the Europeans were called *Pieds-Noirs* ("black feet"). Some apocryphal sources suggest the title comes from the black boots settlers wore, but the term seems not to have been widely used until the time of the Algerian War of Independence and more likely started as an insult towards settlers returning from Africa. In contrast, the vast majority of Muslim Algerians (even veterans of the French army) received neither French citizenship nor the right to vote.

Post-independence

In 1954, the National Liberation Front (FLN) launched the Algerian War of Independence which was a guerrilla campaign. By the end of the war, newly elected President Charles de Gaulle, understanding that the age of empires was ending, held a plebiscite, offering Algerians three options. In a famous speech (4 June 1958 in Algiers) de Gaulle proclaimed in front of a vast crowd of Pieds-Noirs "Je vous ai compris" (I have understood you). Most Pieds-noirs then believed that de Gaulle meant that Algeria would remain French. The poll resulted in a landslide vote for complete independence from France. Over one million people, 10% of the population, then fled the country for France and in just a few months in mid-1962. These included most of the 1,025,000 *Pieds-Noirs*, as well as 81,000 *Harkis* (pro-French Algerians serving in the French Army). In the days preceding the bloody conflict, a group of Algerian Rebels opened fire on a marketplace in Oran killing numerous innocent civilians, mostly women. It is estimated that somewhere between

50,000 and 150,000 *Harkis* and their dependents were killed by the FLN or by lynch mobs in Algeria.

Algeria's first president was the FLN leader Ahmed Ben Bella. He was overthrown by his former ally and defence minister, Houari Boumedienne in 1965. Under Ben Bella the government had already become increasingly socialist and authoritarian, and this trend continued throughout Boumedienne's government. However, Boumedienne relied much more heavily on the army, and reduced the sole legal party to a merely symbolic role. Agriculture was collectivised, and a massive industrialization drive launched. Oil extraction facilities were nationalized. This was especially beneficial to the leadership after the 1973 oil crisis. However, the Algerian economy became increasingly dependent on oil which led to hardship when the price collapsed during the 1980s oil glut.

In foreign policy strained relations with its western neighbour Morocco. Reasons for this include Morocco's disputed claim to portions of western Algeria (which led to the Sand War in 1963), Algeria's support for the Polisario Front for its right to self-determination, and Algeria's hosting of Sahrawi refugees within its borders in the city of Tindouf.

Within Algeria, dissent was rarely tolerated, and the state's control over the media and the outlawing of political parties other than the FLN was cemented in the repressive constitution of 1976.

Boumedienne died in 1978, but the rule of his successor, Chadli Bendjedid, was little more open. The state took on a strongly bureaucratic character and corruption was widespread.

The modernization drive brought considerable demographic changes to Algeria. Village traditions underwent significant change as urbanization increased. New industries emerged and agricultural employment was substantially reduced. Education was extended nationwide, raising the literacy rate from less than 10% to over 60%. There was a dramatic increase in the fertility rate to 7–8 children per mother.

Therefore by 1980, there was a very youthful population and a housing crisis. The new generation struggled to relate to the cultural obsession with the war years and two conflicting protest movements developed: communists, including Berber identity movements; and Islamic 'integrities'. Both groups protested against one-party rule but also clashed with each other in universities and on the streets during the 1980s. Mass protests from both camps in autumn 1988 forced Bendjedid to concede the end of one-party rule.

Algerian Political Events (1991–2002)

Elections were planned to happen in 1991. In December 1991, the Islamic Salvation Front won the first round of the country's first multi-party elections. The military then intervened and cancelled the second round. It forced then-president Bendjedid to resign and banned all political parties based on religion (including the Islamic Salvation Front). A political conflict ensued, leading Algeria into the violent Algerian Civil War.

More than 160,000 people were killed between 17 January 1992 and June 2002. Most of the deaths were between militants and government troops, but a great number of civilians were also killed. The question of who was responsible for these deaths was controversial at the time amongst academic observers; many were claimed by the Armed Islamic Group. Though many of these massacres were carried out by Islamic extremists, the Algerian regime supposedly also used the army and foreign mercenaries to conduct attacks on men, women and children and then proceeded to blame the attacks upon various Islamic groups within the country.

Elections resumed in 1995, and after 1998, the war waned. On 27 April 1999, after a series of short-term leaders representing the military, Abdelaziz Bouteflika, the current president, was chosen by the army.

Post War

By 2002, the main guerrilla groups had either been destroyed or surrendered, taking advantage of an amnesty program, though fighting and terrorism continues in some areas.

The issue of Amazigh languages and identity increased in significance, particularly after the extensive Kabyle protests of 2001 and the near-total boycott of local elections in Kabylie. The government responded with concessions including naming of Tamazight (Berber) as a national language and teaching it in schools. Much of Algeria is now recovering and developing into an emerging economy. The high prices of oil and natural gas are being used by the new government to improve the country's infrastructure and especially improve industry and agricultural land. Recently, overseas investment in Algeria has increased.

Geography

Most of the coastal area is hilly, sometimes even mountainous, and there are a few natural harbours. The area from the coast to the Tell Atlas is fertile. South of the Tell Atlas is a steppe landscape, which ends with the Saharan Atlas; further south, there is the Sahara desert.

The Ahaggar Mountains , also known as the Hoggar, are a highland region in central Sahara, southern Algeria. They are located about 1,500 km (932 mi) south of the capital, Algiers and just west of Tamanghasset.

Algiers, Oran, Constantine, Tizi Ouzou and Annaba are Algeria's main cities.

Tropic of Cancer in the Torrid Zone

In this region even in winter, midday desert temperatures can be very hot. After sunset, however, the clear, dry air permits rapid loss of heat, and the nights are cool to chilly. Enormous daily ranges in temperature are recorded.

The highest temperature recorded in Tiguentour is 140.9 °F (60.5 °C) but this temperature is unofficial and is not recognized by any of the global meteorological organizations. The hottest recognized reading is 135 °F (57.2 °C) at Tindouf. The highest official temperature was 50.6 °C (123.1 °F) at In Salah.

Rainfall is fairly abundant along the coastal part of the Tell Atlas, ranging from 400 to 670 mm (15.7 to 26.4 in) annually, the amount of precipitation increasing from west to east. Precipitation is heaviest in the northern part of eastern Algeria, where it reaches as much as 1,000 mm (39.4 in) in some years.

Farther inland, the rainfall is less plentiful. Prevailing winds that are easterly and north-easterly in summer change to westerly and northerly in winter and carry with them a general increase in precipitation from September through

December, a decrease in the late winter and spring months, and a near absence of rainfall during the summer months. Algeria also has ergs, or sand dunes between mountains, which in the summer time when winds are heavy and gusty, temperatures can get up to 110 °F (43.3 °C).

Politics

The head of state is the President of Algeria, who is elected for a five-year term. The president was formerly limited to two five-year terms but a constitutional amendment passed by the Parliament on November 11, 2008 removed this limitation. Algeria has universal suffrage at 18 years of age. The President is the head of the Council of Ministers and of the High Security Council. He appoints the Prime Minister who is also the head of government. The Prime Minister appoints the Council of Ministers.

The Algerian parliament is bicameral, consisting of a lower chamber, the *National People's Assembly (APN)*, with 380 members; and an upper chamber, the *Council Of Nation*, with 144 members. The APN is elected every five years.

Under the 1976 constitution (as modified 1979, and amended in 1988, 1989, and 1996) Algeria is a multi-party state. The Ministry of the Interior must approve all parties. To date, Algeria has had more than 40 legal political parties. According to the constitution, no political association may be formed if it is "based on differences in religion, language, race, gender or region."

Foreign Relations and Military

The military of Algeria consists of the People's National Army (ANP), the Algerian National Navy (MRA), and the Algerian Air Force (QJJ), plus the Territorial Air Defence Force. It is the direct successor of the Armee de Liberation National (ALN), the armed wing of the nationalist National Liberation Front, which fought French colonial occupation during the Algerian War of Independence (1954–62). The commander-in-chief of the military is the president, who is also Minister of National Defence.

Total military personnel include 147,000 active, 150,000 reserve, and 187,000 paramilitary staff (2008 estimate). Service in the military is compulsory for men aged 19–30, for a total of eighteen months (six training and twelve in civil projects). The total military expenditure in 2006 was estimated variously at 2.7% of GDP (3,096 million), or 3.3% of GDP.

Algeria has its force oriented toward its western neighbour (Morocco) and eastern (Libyan) borders. Its primary military supplier has been the former Soviet Union, which has sold various types of sophisticated equipment under military trade agreements, and the People's Republic of China. Algeria has attempted, in recent years, to diversify its sources of military material. Military forces are supplemented by a 70,000-member gendarmerie or rural police force under the control of the president and 30,000-member *Surete national* or metropolitan police force under the Ministry of the Interior.

In 2007, the Algerian Air Force signed a deal with Russia to purchase 49 MiG-29SMT and 6 MiG-29UBT at an estimated $1.9 billion. They also agreed to return old aircraft purchased from the Former USSR. Russia is also building two 636-type diesel submarines for Algeria.

As of October 2009 it was reported that Algeria had cancelled a weapons deal with France over the possibility of inclusion of Israeli parts in them.

Maghreb Union

Tensions between Algeria and Morocco in relation to the Western Sahara have put great obstacles in the way of tightening the Maghreb Union , which was nominally established in 1989 but carried little practical weight with its coastal neighbours.

Provinces and Districts

Algeria is divided into 48 provinces (*wilayas*), 553 districts (*dairas*) and 1,541 municipalities (*baladiyahs*). Each province, district, and municipality is named after its seat, which is usually the largest city. According to the Algerian constitution, a province is *a territorial collectivity enjoying some economic freedom*.

The People's Provincial Assembly is the political entity governing a province, which has a "president", who is elected by the members of the assembly. They are in turn elected on universal suffrage every five years. The "Wali" (Prefect or governor) directs each province. This person is chosen by the Algerian President to handle the PPA's decisions.

Economy

By far, Algeria's most significant exports today (in terms of financial value) are petroleum and natural gas. The reserves are mostly in the Eastern Sahara; the Algerian government curbed the exports in the 1980s to slow depletion; exports increased again somewhat in the 1990s.

Other significant exports are sheep, oxen, and horses; animal products, such as wool and skins; wine, cereals and seeds, esparto grass, oils and vegetable extracts (chiefly olive oil), iron ore, zinc, natural phosphates, timber, cork, crin vegetal and tobacco. The import of wool exceeds the export. Sugar, coffee, machinery, metal work of all kinds, clothing and pottery are largely imported. Of these by far the greater part comes from France. The British imports consist chiefly of coal, cotton fabrics and machinery.

Algeria trades most extensively with France and Italy, in terms of both imports and exports, but also trades with the United States and Spain. Algeria currently has only one stock exchange, the Algiers Stock Exchange.

Historical Trend

The total imports and exports on the eve of the French invasion (in 1830) did not exceed £175,000. By 1850, the figures had reached £5,000,000; in 1868, £12,000,000; in 1880, £17,000,000; and in 1890, £20,000,000. From this point progress was slower and the figures varied considerably year by year. In 1905 the total value of the foreign trade was £24,500,000. About five-sixths of the trade is with or via France, into which country several Algerian goods have been admitted duty-free since 1851, and all since 1867. French goods, except sugar, have been admitted into Algeria without payment of duty since 1835. After the increase, in 1892, of the French minimum tariff, which applied to Algeria also, foreign trade greatly diminished.

GDP per capita grew 40 percent in the Sixties reaching a peak growth of 538% in the Seventies But this proved unsustainable and growth collapsed to a paltry 9.7% in the turbulent Eighties. Failure of timely reforms by successive governments caused the current GDP per capita to shrink by 28% in the Nineties.

Gross Domestic Product (GDP)

In 2007 Algeria's estimated GDP was US$125.9 billion according to the official exchange rate. Using purchasing power parity, estimated GDP was US$268.9 billion, or US$8,100 on a per capita basis. The estimated real growth rate was 4.6 percent. In 2007 industry accounted for 61 percent of GDP, services constituted 31 percent, and agriculture provided the remaining 8 percent.

Government Budget

In 2007 government revenues of US$58.5 billion exceeded expenditures of US$41.4 billion. Receipts from the hydrocarbons industry usually account for roughly 60 percent of revenues.

Industries

Agriculture, Forestry, and Fishing

Algeria's agricultural sector, which contributes about 8 percent of gross domestic product (GDP) but employs 14 percent of the workforce, is unable to meet the food needs of the country's population. As a result, some 45 percent of food is imported. The primary crops are wheat, barley, and potatoes. Farmers also have had success growing dates for export. Cultivation is concentrated in the fertile coastal plain of the Tell region, which represents just a slice of Algeria's total territory. Altogether, only about 3 percent of Algerian territory is arable. Even in the Tell, rainfall variability has a significant impact on production. Government efforts to stimulate farming in the less arable steppe and desert regions have met with limited success. However, herdsmen maintain livestock, specifically goats, cattle, and sheep, in the High Plateaus region.

Algeria's climate and periodic fires are not conducive to a thriving forestry industry. However, Algeria is a producer of cork and Aleppo pine. In 2005 roundwood removals totalled 7.8 million cubic meters, while sawnwood production amounted to only 13 million cubic meters per year.

Algeria's fishing industry does not take full advantage of the Mediterranean coast, in part because fishing is generally done from small family-owned boats instead of large commercial fishing trawlers. However, the government is attempting to boost the relatively small catch—slightly more than 125,000 metric tons in 2005—by modernizing fishing ports, permitting foreigners to fish in Algerian waters, and subsidizing fishing-related projects.

Fishing

Fishing is a flourishing but minor industry. Fish caught are principally sardines, bonito, mackerel smelt and sprats. Fresh fish are exported to France, dried and preserved fish to Spain and Italy. Coral fisheries are found along the coast from Bona to Tunis. The annual catch averages around 142,000 tons, 54% sardines.

Minerals

Algeria is rich in minerals; the country has many iron, lead, zinc, copper, calamine, antimony and mercury mines. The most productive are those of iron and zinc. Lignite is found in Algiers; immense phosphate beds were discovered near Tebessa in 1891, yielding 313,500 tons in 1905. Phosphate beds are also worked near Setif, Guelma and Ain Beida. There are more than 300 quarries which produce, amongst other stones, onyx and beautiful white and red marbles. Algerian onyx from Ain Tekbalet was used by the Romans, and many ancient quarries have been found near Sidi Ben Yebka, some being certainly those from which the long-lost Numidian marbles were taken. Salt is collected on the margins of the chotts.

Banking and Finance

Algeria's banking sector is dominated by public banks, which suffer from high levels of non-performing loans to state-owned enterprises (SOEs). As of 2007, public banks controlled 95 percent of total bank assets. In 2007 nonperforming loans represented a towering 38 percent of total loans at public banks, according to International Monetary Fund (IMF) estimates. Modest progress has been made in implementing several reforms proposed by the IMF, including replacing bank credits to SOEs with government subsidies; boosting bank supervision, accountability, and transparency; and modernizing the payments system. One specific reform that has been achieved is the establishment in 2006 of the Algerian Real Time Settlements system, which facilitates the prompt and reliable electronic transfer of payments. In November 2007, the proposed sale and privatization of Credit Populaire d'Algerie was postponed because of turbulent market conditions. Recently, HSBC and Deutsche Bank announced that they would commence commercial banking (in the case of HSBC) and investment banking (in the case of Deutsche Bank) in Algeria. Only a few companies are listed on the underdeveloped and relatively opaque Algiers stock exchange.

Tourism

Algeria's tourism industry, which contributes only about 1 percent of GDP, lags behind that of its neighbours Morocco and Tunisia. Algeria receives only about 200,000 tourists and visitors annually. Ethnic Algerian French citizens represent the largest group of tourists, followed by Tunisians. The modest level of tourism is attributable to a combination of poor hotel accommodations and the threat of terrorism. However, the government has adopted a plan known as "Horizon 2025," which is designed to address the lack of infrastructure. Various hotel operators are planning to build hotels, particularly along the Mediterranean coast. Another potential opportunity involves adventure holidays in the south. The Algerian government has set the goal of boosting the number of foreign visitors, including tourists, to 1.2 million by 2010.

Currency, Exchange Rate and Inflation

Algeria's currency is the dinar (DZD). The dinar is loosely linked to the U.S. dollar in a managed float. Algeria's main export, crude oil, is priced in dollars, while most of Algeria's imports are priced in euros. Therefore, the government endeavours to manage fluctuations in the value of the dinar. As of April 2008, US$1 was equivalent to about DZD64.6.

Algeria's foreign currency reserves have grown rapidly since 2000, reflecting rising prices for exported oil. At the end of 2007, foreign reserves totalled US$99.3 billion, up from US$12 billion in 2000 and the equivalent of almost four years of imports.

In 2007 the estimated inflation rate was 4.6 percent.

Labour

The largest employer is government, which claims 32 percent of the workforce. Even though industry is a much larger part of the economy than agriculture, agriculture employs slightly more people (14 percent of the workforce) than industry (13.4 percent of the workforce). One of the reasons for this disparity is that the energy sector is very capital-intensive. Trade accounts for 14.6 percent of the workforce, while the construction and public works sector employs 10 percent, reflecting the government's efforts to upgrade the country's infrastructure and stock of affordable housing.

At the end of 2006, the unemployment rate was about 15.7 percent, but the rate among those under the age of 25 was 70 percent. In 2005 the labour participation rate was only 52 percent, versus an Organisation for Economic Co-operation and Development average of 70 percent. New entrants to the workforce and the lack of emigration options make unemployment a chronic problem and an important challenge to the government. Given its highly capital-intensive nature, the hydrocarbons industry is not in a position to employ many job seekers.

Foreign Economic Relations

In its foreign economic relations, Algeria is seeking more trade and foreign investment. For example, Algeria's hydrocarbons law passed in April 2005 is designed to encourage foreign investment in energy exploration. Increased production could raise Algeria's profile as a member of the Organization of the Petroleum Exporting Countries. In keeping with its pro-trade agenda, Algeria achieved association status with the European Union (EU) in September 2005. Over a 12-year period, the association agreement is expected to enable Algeria to export goods to the EU tariff-free, while it gradually lifts tariffs on imports from the EU. Algeria has signed bilateral investment agreements with 20 different nations, including many European countries, China, Egypt, Malaysia, and Yemen. In July 2001, the United States and Algeria agreed on a framework for discussions leading to such an agreement, but a final treaty has not yet been negotiated. Ultimately, trade liberalization, customs modernization, deregulation, and banking reform are designed to improve the country's negotiating position as it seeks accession to the World Trade Organization. In 2007 Algerian imports totalled US$26.08 billion. The principal imports were capital goods, foodstuffs, and consumer goods. The top import partners were France (22 percent), Italy (8.6 percent), China (8.5 percent), Germany (5.9 percent), Spain (5.9 percent), the United States (4.8 percent), and Turkey (4.5 percent).

In 2007 Algeria exported US$63.3 billion, more than twice as much as it imported. Exports accounted for 30 percent of gross domestic product (GDP). Hydrocarbon products constituted at least 95 percent of export earnings. The principal exports were petroleum, natural gas, and petroleum products. The top export partners were the United States (27.2 percent), Italy (17 percent), Spain (9.7 percent), France (8.8 percent), Canada (8.1 percent), and Belgium (4.3 percent). Algeria supplies 25 percent of the European Union's natural gas imports. In 2007 Algeria posted a positive merchandise trade balance of US$37.2 billion. In 2007 Algeria achieved a positive current account balance of US$31.5 billion. High prices for Algeria's energy exports are the main driver for the improvement in the current account balance. Reflecting strong oil export revenues, external debt is on a downward trajectory. For example, these revenues facilitated early repayments of US$900 million in loans from the African Development Bank and Saudi Arabia. In March 2006, Algeria's purchase of 78 aircraft from Russia led to the cancellation of Algeria's entire debt to Russia. In 2006 external debt was estimated at US$4.4 billion, down from US$23.5 billion in 2003.

In 2006 foreign direct investment (FDI) in Algeria totalled US$1.8 billion. The petrochemical, transport, and utilities sectors have been recent beneficiaries of FDI. FDI into the oil sector was expected to rise as a result of a hydrocarbons law, approved in April 2005, that created a more even playing field for foreign oil companies to compete with Algeria's state-owned oil company, Sonatrach, for exploration and production contracts. Algeria also is seeking foreign investment in power and water systems.

As of August 2006, cumulative World Bank assistance to Algeria totalled US$5.9 billion, encompassing 72 projects. Currently, the World Bank is pursuing seven projects, specifically budget modernization, mortgage finance, natural disaster recovery, energy and mining, rural employment, telecommunications, and transportation. In 2005 economic assistance to Algeria from the United States amounted to US$4.4 million, most of which was attributable to the Middle East Partnership Initiative (MEPI) and the remainder to International Military Education and Training (IMET). MEPI encourages economic, political, and educational reform in the Middle East. In 2006 IMET, which provides U.S. military training to foreign troops, had a budget of US$823 million. In 2005 the European Union contributed US$58 million to Algeria's economic development under the Euro-Mediterranean Partnership. The fossil fuels energy sector is the backbone of Algeria's economy, accounting for roughly 60% of budget revenues, 30% of GDP, and over 95% of export earnings. The country ranks fourteenth in petroleum reserves, containing 11.8 billion barrels of proven oil reserves with estimates suggesting that the actual amount is even more. The U.S. Energy Information Administration reported that in 2005, Algeria had 160 trillion cubic feet ($4.5\times10^{\Delta12}\ m^3$) of proven natural gas reserves, the eighth largest in the world.

Algeria's financial and economic indicators improved during the mid-1990s, in part because of policy reforms supported by the International Monetary Fund (IMF) and debt rescheduling from the Paris Club. Algeria's finances in 2000 and 2001 benefited from an increase in oil prices and the government's tight fiscal policy, leading to a large increase in the trade surplus, record highs in foreign exchange reserves, and reduction in foreign debt.

The government's continued efforts to diversify the economy by attracting foreign and domestic investment outside the energy sector have had little success in reducing high unemployment and improving living standards, however. In 2001, the government signed an Association Treaty with the European Union that will eventually lower tariffs and increase trade. In March 2006, Russia agreed to erase $4.74 billion of Algeria's Soviet-era debt during a visit by President Vladimir Putin to the country, the first by a Russian leader in half a century. In return, president Bouteflika agreed to buy $7.5 billion worth of combat planes, air-defence systems and other arms from

Russia, according to the head of Russia's state arms exporter Rosobor on export. Algeria also decided in 2006 to pay off its full $8bn (£4.3bn) debt to the Paris Club group of rich creditor nations before schedule. This will reduce the Algerian foreign debt to less than $5bn in the end of 2006. The Paris Club said the move reflected Algeria's economic recovery in recent years.

Demographics

The population of Algeria is 35,190,000 (January 2009 est.), with 99% classified ethnically as Arab or Berber. About 90% of Algerians live in the northern, coastal area; the minority who inhabit the Sahara are mainly concentrated in oases, although some 1.5 million remain nomadic or partly nomadic. Almost 30% of Algerians are under the age of 15. Algeria has the fourth lowest fertility rate in the Greater Middle East, after those of Cyprus, Tunisia, and Turkey.

The ethnic ancestry of most Algerians is composed of Berbers (mostly Zenatas and Numidians) and Middle-East populations who settled Northwest Africa at various period of the history such as ancient Punics and Arabs (such as Banu Hilal tribes).

Linguistically, ~83% of Algerians speak Algerian Arabic, while Berber dialects are spoken in the Kabyle and Chaoui regions. French is widely understood, and Standard Arabic (FosHaa) is taught to and understood by most Algerian youth.

Europeans account for less than 1% of the population, inhabiting almost exclusively the largest metropolitan areas. However, during the colonial period there was a large (15.2% in 1962) European population, consisting primarily of French people, in addition to Spaniards in the west of the country, Italians and Maltese in the east, and other Europeans such as Greeks in smaller numbers. Known as *pieds-noirs*, European colonists were concentrated on the coast and formed a majority of the population of Oran (60%) and important proportions in other large cities like Algiers and Annaba. Almost all of this population left during or immediately after the country's independence from France.

Shortages of housing and medicine continue to be pressing problems in Algeria. Failing infrastructure and the continued influx of people from rural to urban areas has overtaxed both systems. According to the UNDP, Algeria has one of the world's highest per housing unit occupancy rates for housing, and government officials have publicly stated that the country has an immediate shortfall of 1.5 million housing units.

Women make up 70 percent of Algeria's lawyers and 60 percent of its judges, and also dominate the field of medicine. Increasingly, women are contributing more to household income than men. Sixty percent of university students are women, according to university researchers.

It is estimated that 95,700 refugees and asylum-seekers have sought refuge in Algeria. This includes roughly 90,000 from Morocco and 4,100 from Palestine. An estimated 90,000 to 160,000 Sahrawis from Western Sahara live in refugee camps in the Algerian part of the Sahara Desert. There are currently around 35,000 Chinese migrant workers in Algeria.

Ethnic Groups

Ethnic Arabs compose Algeria's majority while a significant Berber population make up the rest of Algeria's ethnic groups. No official figures can be given, because Algerian law forbids population censuses based on ethnic, religious and linguistic criteria. Ethnic Berbers are divided into several groups including Kabyle in the mountainous north-central area, Chaoui in the eastern Atlas Mountains among other groups.

A few thousand French from the colonial era remain in the country after nearly the entire Franco-Algerian community fled out after the war of Independence in the 1950s, and smaller numbers of ethnic Europeans, as well other Arab nationalities and Sub-Saharan Africans reside in the country.

Languages

Arabic: Algerian colloquial Arabic is spoken as a native language by more than 83% of the population; of these, over 78% speak Algerian Arabic and around 5% Hassaniya. Algerian Arabic is spoken as a second language by many Berbers. However, in the media and on official occasions the spoken language is Standard Arabic.

Berber: Berber languages are spoken by 15% of Algeria's population mainly in Kabylia, in the Acres, and in the Sahara (by Tuaregs). Until the Phoenicians' arrival, Berber was spoken throughout Algeria, as later attested by early Tifinagh inscriptions. Despite the growth of Punic, Latin and later, Arabic, Berber remained the main language of Algeria until the invasion of the populous Banu Hilal and Banu Sulaym tribal confederation in the 11th century. The Berbers (or Imazighen) speak one of the various dialects of Tamazight, which add up to around 28% of the population. Arabic remains Algeria's only official language, although Tamazight has recently been recognized as a national language.

French: French is the most widely studied foreign language in the country, and a majority of Algerians can understand it and speak it, though it is usually not spoken in daily life. Since independence, the government has pursued a policy of linguistic Arabization of education and bureaucracy, which has resulted in limiting the use of Berber and the Arabization of many Berber-speakers. The strong position of French in Algeria was little affected by the Arabization policy. All scientific and business university courses are still taught in French. Recently, schools have begun to incorporate French into the curriculum as early as children are taught written classical Arabic. French is also used in media and business. After a political debate in Algeria in the late 1990s about whether to replace French with English in the educational system, the government decided to retain French. English is taught in the first year of middle schools. The Punic tongue, a Phoenician language, is believed to have been spoken in several areas of the country. The Punic language died out after the 6th century AD. Algerian cities have commonly been given Punic and ancient Roman names.

Religion

Islam is the predominant religion, followed by more than 99 percent of the country's population. This figure includes all these born in families considered of Muslim descent. Officially, nearly 100% of all Algerians are Muslims. Almost all Algerians follow Sunni Islam, with the exception of some 200,000 ibadis in the M'zab Valley in the region of Ghardaia.

There are also some 150,000 Christians in the country, including about 10,000 Roman Catholics and 50,000 to 100,000 evangelical Protestants (mainly Pentecostal), according to the Protestant Church of Algeria's leader Mustapha Krim. Some sources claim more than one million Christians in Algeria, most of them live in the Kabyl area where there are more than 70 underground churches.

Algeria had an important Jewish community until the 1960s. Nearly all of this community emigrated following the country's independence, although a very small number of Jews continue to live in Algiers.

The nation has experienced a decline in Christianity as a result of Islamization for over a millennia. Catholic French missionaries converted some Muslim Algerians, and their descendants are called *Evolves*, who moved on to France in Europe.

Health

In 2002 Algeria had inadequate numbers of physicians (1.13 per 1,000 people), nurses (2.23 per 1,000 people), and dentists (0.31 per 1,000 people). Access to "improved water sources" was limited to 92 percent of the population in urban areas and 80 percent of the population in rural areas. Some 99 percent of Algerians living in urban areas, but only 82 percent of those living in rural areas, had access to "improved sanitation."

According to the World Bank, Algeria is making progress toward its goal of "reducing by half the number of people without sustainable access to improved drinking water and basic sanitation by 2015." Given Algeria's young population, policy favours preventive health care and clinics over hospitals. In keeping with this policy, the government maintains an immunization program. However, poor sanitation and unclean water still cause tuberculosis, hepatitis, measles, typhoid fever, cholera, and dysentery. The poor generally receive health care free of charge.

Education

Education is officially compulsory for children between the ages of six and fifteen. In the year 1997, there was an outstanding amount of teachers and students in primary schools. Approximately 5% of the adult population of the country is illiterate. In Algeria there are 43 universities, 10 colleges, and

7 institutes for higher learning. The University of Algiers (founded in 1909) has about 267,142 students. The Algerian school system is structured into Basic, General Secondary, and Technical Secondary levels:

Basic; Ecole fondamentale (Fundamental School), Length of program: nine years, Age range: six to fifteen, Certificate/ diploma awarded: Brevet d'Enseignement Moyen B.E.M.

General Secondary: Lycee d'Enseignement general (School of General Teaching), lycees polyvalents (General-Purpose School), Length of program: three years, Age range: 15 to 18, Certificate/diploma awarded: Baccalaureat de l'Enseignement secondaire, (Bachelor's Degree of Secondary School).

Technical Secondary: Lycees d'Enseignement technique (Technical School), Length of program: three years, Certificate/ diploma awarded: Baccalaureat technique (Technical Bachelor's Degree).

Culture

Modern Algerian literature, split between Arabic and French, has been strongly influenced by the country's recent history. Famous novelists of the twentieth century include Mohammed Dib, Albert Camus, and Kateb Yacine, while Assia Djebar is widely translated. Among the important novelists of the 1980s were Rachid Mimouni, later vice-president of Amnesty International, and Tahar Djaout, murdered by an Islamist group in 1993 for his secularist views.

In philosophy and the humanities, Jacques Derrida, the father of deconstruction, was born in El Biar in Algiers; Malek Bennabi and Frantz Fanon are noted for their thoughts on decolonization; Augustine of Hippo was born in Tagaste (modern-day Souk Ahras); and Ibn Khaldun, though born in Tunis, wrote the Muqaddima while staying in Algeria. Algerian culture has been strongly influenced by Islam, the main religion. The works of the Sanusi family in pre-colonial times, and of Emir Abdelkader and Sheikh Ben Badis in colonial times, are widely noted. The Latin author Apuleius was born in Madaurus (Mdaourouch), in what later became Algeria.

12

Angola

Angola, officially the Republic of Angola, is a country in south-central Africa bordered by Namibia on the south, Democratic Republic of the Congo on the north, and Zambia on the east; its west coast is on the Atlantic Ocean with Luanda as its capital city. The exclave province of Cabinda has a border with the Republic of the Congo and the Democratic Republic of the Congo.

Angola was a Portuguese overseas territory from the 16th century to 1975. After independence, Angola was the scene of an intense civil war from 1975 to 2002. The country is the second-largest petroleum and diamond producer in sub-Saharan Africa; however, its life expectancy and infant mortality rates are both among the worst ranked in the world. In August 2006, a peace treaty was signed with a faction of the FLEC, a separatist guerrilla group from the Cabinda exclave in the North, which is still active. About 65% of Angola's petroleum comes from that region.

Geography of Angola

Angola is located on the western Atlantic Coast of southern Africa between Namibia and the Republic of the Congo. It also is bordered by the Democratic Republic of the Congo and Zambia to the east. The country consists of a sparsely watered and somewhat sterile coastal plain extending inland for a distance varying from 30 to 100 miles (48 to 160 km). Slightly inland and parallel to the coast is a belt of hills and mountains and behind those a large plateau.

Geology

The rock formations of Angola are met with in three distinct regions:

1. the littoral zone,
2. the median zone formed by a series of hills more or less parallel with the coast,
3. the central plateau.

The central plateau consists of ancient crystalline rocks with granites overlain by non-fossiliferous sandstones and conglomerates of Paleozoic age. The outcrops are largely hidden under laterite.

The median zone is composed largely of crystalline rocks with granites and some Palaeozoic unfossiliferous rocks. The littoral zone contains the only fossiliferous strata. These are of Tertiary and Cretaceous ages, the latter rocks resting on a reddish sandstone of older date.

The Cretaceous rocks of the Dombe Grande region (near Benguella) are of Albian age and belong to the *Acanthoceras mamillari* zone. The beds containing *Schloenbachia inflata* are referable to the Gault. Rocks of Tertiary age are met with at Dombe Grande, Mossamedes and near Loanda.

The sandstones with gypsum, copper and sulfur of Dombe are doubtfully considered to be of Triassic age. Recent eruptive rocks, mainly basalts, form a line of hills almost bare of vegetation between Benguella and Mossamedes. Nepheline basalts and liparites occur at Dombe Grande. The presence of gum copal in considerable quantities in the superficial rocks is characteristic of certain regions.

Location

Southern Africa, bordering the South Atlantic Ocean, between Namibia and Democratic Republic of the Congo.

Area

- *total:* 1,246,700 km^2 (481,354 sq mi)
- *land:* 1,246,700 km^2 (481,354 sq mi)
- *water:* 0 km^2 (0 sq mi).

Country comparison to the world: 30

Area Comparative

- Australia comparative: smaller than the Northern Territory
- Canada comparative: slightly smaller than the Northwest Territories
- United Kingdom comparative: 5 times bigger than the UK
- United States comparative: slightly less than twice the size of Texas.

Capital

- Luanda (Sao Paulo de Loanda) - port - railhead.

Major Cities

- Amboim (Porto Amboim)
- Bailundo (Vila Teixeira da Silva)
- Benguela (Sao Felipe de Benguella) - port - railhead
- Caala (Vila Robert Williams)
- Calandula (Duque de Braganca)
- Camacupa (Vila General Machado)
- Chibia (Vila Joao de Almeida)
- Ganda (Vila Mariano Machado)
- Huambo (Nova Lisboa) - rail
- Kuito (Silva Porto)
- Kuvango (Vila da Ponte)
- Lubango (Sa da Bandeira)
- Luena (Vila Luso)
- Massango (Forte República)
- Mbanza Congo (Sao Salvador do Congo)
- Menongue (Serpa Pinto) - railhead
- Namibe (Mocamedes) - port - railhead
- N'Dalatando (Vila Salazar) - rail
- N'Giva (Vila Pereira d'Eca)
- Saurimo (Vila Henrique de Carvalho)
- Soyo (Santo Antonio do Zaire)
- Sumbe (Novo Redondo)

- Tombua (Porto Alexandre)
- Uije (Carmona)
- Other Towns in Angola.

Land Boundaries

- total: 5,198 km
- border countries: Democratic Republic of the Congo 2,511 km (of which 220 km is the boundary of discontiguous Cabinda Province), Republic of the Congo 201 km, Namibia 1,376 km, Zambia 1,110 km.

Coastline: 1,600 km.

Maritime claims:

- contiguous zone: 24 nmi (44 km)
- exclusive economic zone: 200 nmi (370 km)
- territorial sea: 12 nmi (22 km).

Climate

Like the rest of tropical Africa, Angola experiences distinct, alternating rainy and dry seasons. The coastal strip is tempered by the cool Benguela Current, resulting in a climate similar to coastal Peru or Baja California. It is semiarid in South and along coast to Luanda.

There is a short rainy season lasting from February to April. Summers are hot and dry, while winters are mild. North has cool, dry season (May to October) and hot, rainy season (November to April). In the interior, above 3,300 ft (1,000 m), the temperature and rainfall decrease.

The interior highlands have a mild climate with a rainy season from November through April followed by a cool dry season from May to October. The plateau climate is healthy and invigorating. The mean annual temperature at Sao Salvador do Congo is 22.2 °C (72.0 °F); at Luanda, 23.3 °C (73.9 °F); and at Caconda, {convert|19.5|C|F}}. The climate is greatly influenced by the prevailing winds, which arc W., S.W. and S.S.W. Two seasons are distinguished - the cool, from June to September; and the rainy, from October to May. The heaviest rainfall occurs in April, and is accompanied by violent storms. The far north and Cabinda enjoy rain throughout much of the year.

Terrain

Angola has four principal natural regions: the arid coastal lowland, stretching from Namibia to Luanda and characterized by low plains and terraces; green hills and mountains, rising inland from the coast into a great escarpment; a large area of high inland plains of dry savanna, called the high plateau (planalto), which extends eastward and south-east from the escarpment; and rain forest in the north and in Çabinda. The highest point in Angola is Morro de Môco, at 2,620 m. Elevations generally range from 3,000 to 6,000 feet (910 to 1,800 m).

Coastal Lowland

The coast is for the most part flat, with occasional low cliffs and bluffs of red sandstone. There is but one deep inlet of the sea - Great Fish Bay (or Baia dos Tigres). Farther north are Port Alexander, Little Fish Bay and Lobito Bay, while shallower bays are numerous. Lobito Bay has water sufficient to allow large ships to unload close inshore. The coastal lowland rises from the sea in a series of low terraces. This region varies in width from about 25 kilometres near Benguela to more than 150 kilometres in the Cuanza River Valley just south of Angola's capital, Luanda, and is markedly different from Angola's highland mass. The Atlantic Ocean's cold, northward flowing Benguela Current substantially reduces precipitation along the coast, making the region relatively arid or nearly so south of Benguela (where it forms the northern extension of the Namib Desert), and quite dry even in its northern reaches. Even where, as around Luanda, the average annual rainfall may be as much as fifty centimetres, it is not common for the rains to fail. Given this pattern of precipitation, the far south is marked by sand dunes, which give way to dry scrub along the middle coast. Portions of the northern coastal plain are covered by thick brush.

Hills and Mountains

The approach to the great central plateau of Africa is marked by the west-central highlands, a series of irregular escarpments and cuestas parallel to the coast at distances ranging from 20 kilometres to 100 kilometres inland as Tala Mugongo (4,400 ft/1,300 m), Chella and Vissecua (5,250 or 6,500 ft/1,600 or 2,000 m). The Cuanza River divides the mountain zone into two parts. The northern part rises gradually from the coastal zone to an average elevation of 500 meters, with crests as high as 1,000 meters to 1,800 meters. South of the Cuanza River, the hills rise sharply from the coastal lowlands and form a high escarpment, extending from a point east of Luanda and running south through Namibia. The highest peak is Mount Moco (2,620 m/8,600 ft), and the escarpment is steepest in the far south in the Serra da Chella mountain range. In Benguela Province other high points are Loviti (7,780 ft/ 2,370 m), in 12° 5' S., and Mt. Elonga (7,550 ft/2,300 m). South of the Cuanza is the volcanic mountain Caculo-Cabaza (3,300 ft/ 1,000 m).

High Plateau

The high plateau, with an altitude ranging from 4,000 to 6,000 feet (1,200 to 1,800 m), lies to the east of the hills and mountains and dominates Angola's terrain. It consists of well-watered, wide, rolling plains, and low hills with scanty vegetation. The surface of the plateau is typically flat or rolling, but parts of the Benguela Plateau and the Humpata Highland area of the Huila Plateau in the south reach heights of 2,500 meters and more. The Malanje Plateau to the north rarely exceeds 1,000 meters in height. The Benguela Plateau and the coastal area in the immediate environs of Benguela and Lobito, the Bie Plateau, the Malanje Plateau, and a small section of the Huila Plateau near the town of Lubango have long been among the most densely settled areas in Angola. In the east the tableland falls away to the basins of the Congo and Zambezi, to the south in Namibia it merges into a barren sandy desert.

Drainage

The Zambezi River and several tributaries of the Congo River have their sources in Angola. A large number of rivers originate in the central uplands, but their patterns of flow are diverse and their ultimate outlets varied. A number flow in a more or less westerly course to the Atlantic Ocean, providing water for irrigation in the dry coastal strip and the potential for hydroelectric power, only some of which had been realized by 1988. Two of Angola's most important rivers, the Cuanza (Kwanza) and the Cunene (Kunene), take a more indirect route to the Atlantic, the Cuanza flowing north and the Cunene flowing south before turning west. The Cuanza is the only river wholly within Angola that is navigable—for nearly 200 kilometres from its mouth- -by boats of commercially or militarily significant size. The Congo River, whose mouth and western end form a small portion of Angola's northern border with Zaire, is also navigable.

North of the Lunda Divide the Kwango and many other streams flow north from the tableland to join the Kasai River (one of the largest affluents of the Congo), which in its upper course forms for fully 300 mi (480 km) the boundary between Angola and the Congo. South of the divide some rivers flow into the Zambezi River system and thence to the Indian Ocean, others to the Okavango River (as the Cubango River is called along the border with Namibia and in Botswana) and thence to Lake Ngami and the Okavango Swamp in Botswana. The tributaries of the Cubango River and several of the southern rivers flowing to the Atlantic are seasonal, completely dry much of the year.

Land use and Hazards

Natural resources: petroleum, diamonds, iron ore, phosphates, copper, feldspar, gold, bauxite, uranium.

Land use:

- arable land: 2.65%
- permanent crops: 0.23%
- other: 97.12% (2005).

Irrigated land: 800 km^2 (2003 est.)

Total renewable water resources: 184 km^3 (44.1 cu mi) (1987).

Freshwater withdrawal (domestic/industrial/agricultural)

- *total:* 0.35 km^3 (0.084 cu mi)/yr (23%/17%/60%)
- *per capita:* 22 m^3 (29 cu yd)/yr (2000).

Natural hazards: locally heavy rainfall causes periodic flooding on the plateau.

Environment - Current Issues

Overuse of pastures and subsequent soil erosion attributable to population pressures; desertification; deforestation of tropical rain forest, in response to both international demand for tropical timber and to domestic use as fuel, resulting in loss of biodiversity; soil erosion contributing to water pollution and silting of rivers and dams; inadequate supplies of potable water

Environment - international agreements:

- *party to:* Biodiversity, Climate Change, Desertification, Law of the Sea, Marine Dumping, Ozone Layer Protection, Ship Pollution (MARPOL 73/78)
- *signed, but not ratified:* none of the selected agreements

Flora and Fauna

Both flora and fauna are those characteristic of the greater part of tropical Africa. As far south as Benguela the coast region is rich in oil palms and mangroves. In the Northern part of the province are dense forests. In the South towards the Kunene are regions of dense thorn scrub. Rubber vines and trees are abundant, but in some districts their number has been considerably reduced by the primitive methods adopted by native collectors of rubber.

The species most common are various root rubbers, notably the *Carpodinus chylorrhiza*. This species and other varieties of carpodinus are very widely distributed. Landolphias are also found. The coffee, cotton and Guinea pepper plants are indigenous, and the tobacco plant flourishes in several districts. Among the trees are several which yield excellent timber, such as the tacula (*Pterocarpus tinctorius*), which grows to an immense size, its wood being blood-red in colour, and the Angola mahogany. The bark of the musuemba (*Albizzia coriaria*) is largely used in the tanning of leather. The mulundo bears a fruit about the size of a cricket ball covered with a hard green shell and containing scarlet pips like a pomegranate.

The fauna includes the lion, leopard, cheetah, elephant, giraffe, rhinoceros, hippopotamus, buffalo, zebra, kudu and many other kinds of antelope, wildpig, ostrich and crocodile. Angola previously served as a habitat for the endangered African Wild Dog, which is now deemed to be extinct within the entire country, stemming from human activities during the period 1965 to 1991. Among fish are the barbel, bream and African yellow fish.

Ecoregions of Angola

The following ecoregions have been described in Angola:

- Angolan Scarp savanna and woodlands, the steep coastal escarpment.
- Angolan montane forest-grassland mosaic, the inland slopes of the central highlands which are covered in grassland and contain the remaining patches of mountain woodland;
- Angolan Miombo woodlands, much of the large inland plain, indeed most of central Angola.
- Angolan Mopane woodlands, an area in the south, mostly comprising Cunene Province and extending across the border into neighbouring Namibia.

Geography - note: the province of Cabinda is an exclave, separated from the rest of the country by the Democratic Republic of the Congo

Extreme points

This is a list of the extreme points of Angola, the points that are farther north, south, east or west than any other location.

Angola;

- Northernmost Point - unnamed point on the border with Republic of the Congo (north of the town Caio Bemba, Cabinda province (an Angolan exclave)
- Easternmost Point - unnamed location on a river section of the border with Zambia (north of the town Sapeta in Zambia), Moxico province
- Southern most Point - on the point where the Cunene River section of the border with Namibia terminates at the Caprivi Strip (immediately north of the town Andara in Namibia, Cuando Cubango province
- Western most Point - Baia dos Tigres island, Namibe Province.

Angola (mainland);

- Northernmost Point - a point on the border with the Democratic Republic of Congo immediately to the north-west of the town Luvo, Zaire Province

- Easternmost Point - unnamed point on a river section of the border with Zambia (north of the town Sapeta in Zambia), Moxico province
- Southern most Point - on the point where the Cunene River section of the border with Namibia terminates at the Caprivi Strip (immediately north of the town Andara in Namibia, Cuando Cubango province
- Western most Point - unnamed headland west of Tombua (Porto Alexandre), Namibe.

Climate

Angola's average temperature on the coast is 60 °F (16 °C) in the winter and 70 °F (21 °C) in the summer. It has two seasons; dry season (May to October) and hot rainy season (November to April).

Economy

Angola's economy has undergone a period of transformation in recent years, moving from the disarray caused by a quarter century of civil war to being the fastest growing economy in Africa and one of the fastest in the world. In 2004, China's Eximbank approved a $2 billion line of credit to Angola. The loan is being used to rebuild Angola's infrastructure, and has also limited the influence of the International Monetary Fund in the country.

Growth is almost entirely driven by rising oil production which surpassed 1.4 million barrels per day (220,000 m³/d) in late-2005 and was expected to grow to 2 million barrels per day (320,000 m³/d) by 2007. Control of the oil industry is consolidated in Sonangol Group, a conglomerate which is owned by the Angolan government. In December 2006, Angola was admitted as a member of OPEC. The economy grew 18% in 2005, 26% in 2006 and 17.6% in 2007 and it's expected to stay above 10% for the rest of the decade. The security brought about by the 2002 peace settlement has led to the resettlement of 4 million displaced persons, thus resulting in large-scale increases in agriculture production.

The country's economy has grown since achieving political stability in 2002. However, it faces huge social and economic problems as a result of the almost continual state of conflict from 1961 onwards, although the highest level of destruction and socio-economic damage took place after the 1975 independence, during the long years of civil war. The oil sector, with its fast-rising earnings has been the main driving force behind improvements in overall economic activity – nevertheless, poverty remains widespread. Anti-corruption watchdog Transparency International rated Angola one of the 10 most corrupt countries in the world in 2005. The capital city is the most developed and the only large economic centre worth mentioning in the country, however, slums called *musseques*, stretch for miles beyond Luanda's former city limits.

According to the Heritage Foundation, a conservative American think tank, oil production from Angola has increased so significantly that Angola now is China's biggest supplier of oil.

Before independence in 1975, Angola was a breadbasket of southern Africa and a major exporter of bananas, coffee and sisal, but three decades of civil war (1975–2002) destroyed the fertile countryside, leaving it littered with landmines and driving millions into the cities. The country now depends on expensive food imports, mainly from South Africa and Portugal, while more than 90 percent of farming is done at family and subsistence level. Thousands of Angolan small-scale farmers are trapped in poverty.

Transport

Transport in Angola consists of:

- Three separate railway systems totalling 2,761 km
- 76,626 km (47,613 mi) of highway of which 19,156 km (11,903 mi) is paved
- 1,295 navigable inland waterways
- Eight major sea ports
- 243 airports, of which 32 are paved.

Travel on highways outside of towns and cities in Angola (and in some cases within) is often not best advised for those without four-by-four vehicles. Whilst a reasonable road infrastructure has existed within Angola, time and the war have taken their toll on the road surfaces, leaving many severely potholed, littered with broken asphalt. In many areas drivers have established alternate tracks to avoid the worst parts of the surface, although careful attention must be paid to the presence or absence of landmine warning markers by the side of the road. The Angolan government has contracted the restoration of many of the country's roads. The road between Lubango and Namibe, for example, was completed recently with funding from the European Union, and is comparable to many European main routes. Progress to complete the road infrastructure is likely to take some decades, but substantial efforts are already being made in the right directions.

Demographics

Angola's population is estimated to be 12,799,293 (July 2009). It is composed of Ovimbundu 37%, Mbundu 25%, Bakongo 13%, *mesticos* (mixed European and native African) 2%, European 1%, and 22% 'other' ethnic groups. The two Mbundu and Ovimbundu nations combined form a majority of the population, at 62%.

It is estimated that Angola was host to 12,100 refugees and 2,900 asylum seekers by the end of 2007. 11,400 of those refugees were originally from the Democratic Republic of Congo (Congo-Kinshasa) who arrived in the 1970s. As of 2008 there were an estimated 400,000 DRC migrant workers, at least 30,000 Portuguese, and 100,000+ Chinese living in Angola. Prior to independence in 1975, Angola had a community of approximately 500,000 Portuguese.

Languages

Portuguese is spoken as a first language by 80% of the population, and as a second language by another 20%. The dominance of Portuguese over the native Kimbundu and other African languages is due to a strong influence from Portugal, as opposed to in Mozambique, which being more remote from the Lusosphere, retained a majority of Bantu language speakers.

Religion

Christianity is the major religion in Angola. The World Christian Database states that the Angolan population is 93.5% Catholic, 4.7% ethnoreligionist (indigenous), 0.6% Muslim, 0.9% Agnostic and 0.2% non-religious. However, other sources put the percent of Christians at 53% with the remaining population adhering to indigenous beliefs. According to these sources, of Christians in Angola, 72% are Roman Catholic, and the 28% are divided among the Baptist, Presbyterian, Reformed Evangelical, Pentecostal, Methodist and other Christian denominations.

In a study assessing nations' levels of religious regulation and persecution with scores ranging from 0–10 where 0 represented low levels of regulation or persecution, Angola was scored 0.8 on Government Regulation of Religion, 4.0 on Social Regulation of Religion, 0 on Government Favouritism of Religion and 0 on Religious Persecution.

The largest Protestant denominations include the Methodists, Baptists, Congregationalists (United Church of Christ), and Assemblies of God.

The largest syncretic religious group is the Kimbanguist Church, whose followers believe that a mid-20th century Congolese pastor named Joseph Kimbangu was a prophet. A small portion of the country's rural population practices animism or traditional indigenous religions. There is a small Islamic community based on migrants from West Africa.

In colonial times, the country's coastal populations primarily were Catholic while the Protestant mission groups were active inland. With the massive social displacement caused by 26 years of civil war, this rough division is no longer valid.

Foreign missionaries were very active prior to independence in 1975, although the Portuguese colonial authorities expelled many Protestant missionaries and closed mission stations based on the belief that the missionaries were inciting pro-independence sentiments. Missionaries have been able to return to the country since the early 1990s, although security conditions due to the civil war have prevented them from restoring many of their former inland mission stations.

The Roman Catholic denomination mostly keeps to itself in contrast to the major Protestant denominations which more actively proselytize. The major Protestant denominations provide help for the poor in the form of crop seeds, farm animals, medical care and education in the English language, math, history and religion.

Health

A 2007 survey concluded that low and deficient niacin status was common in Angola. Epidemics of cholera, malaria, rabies and African hemorrhagic fevers like Marburg hemorrhagic fever, are common diseases in several parts of the country. Many regions in this country have high incidence rates of tuberculosis and high HIV prevalence rates. Dengue, filariasis, leishmaniasis, and onchocerciasis (river blindness) are other diseases carried by insects that also occur in the region. Angola has one of the highest infant mortality rates in the world and the world's 2nd lowest life expectancies.

Education

Although by law, education in Angola is compulsory and free for 8 years, the government reports that a certain percentage of students are not attending school due to a lack of school buildings and teachers. Students are often responsible for paying additional school-related expenses, including fees for books and supplies.

In 1999, the gross primary enrollment rate was 74 percent and in 1998, the most recent year for which data are available, the net primary enrollment rate was 61 percent. Gross and net enrollment ratios are based on the number of students formally registered in primary school and therefore do not necessarily reflect actual school attendance. There continue to be significant disparities in enrollment between rural and urban areas. In 1995, 71.2 percent of children ages 7 to 14 years were attending school. It is reported that higher percentages of boys attend school than girls. During the Angolan Civil War (1975–2002), nearly half of all schools were reportedly looted and destroyed, leading to current problems with overcrowding.

The Ministry of Education hired 20,000 new teachers in 2005, and continued to implement teacher trainings. Teachers tend to be underpaid, inadequately trained, and overworked (sometimes teaching two or three shifts a day). Some teachers may also reportedly demand payment or bribes directly from their students. Other factors, such as the presence of landmines, lack of resources and identity papers, and poor health also prevent children from regularly attending school. Although budgetary allocations for education were increased in 2004, the education system in Angola continues to be extremely underfunded.

Literacy is quite low, with 67.4% of the population over the age of 15 able to read and write in Portuguese. 82.9% of males and 54.2% of women are literate as of 2001. Since independence from Portugal in 1975, a number of Angolan students continued to be admitted every year at high schools, polytechnical institutes, and universities Portuguese, Brazilian and Cuban through bilateral agreements; in general these students belong to the Angolan elites.

Culture

Portugal ruled over Angola for 400 years and both countries share cultural aspects: language (Portuguese) and main religion (Roman Catholic Christianity). However, the Angolan culture is mostly native Bantu which was mixed with Portuguese culture. The diverse ethnic communities with their own cultural traits, traditions and native languages or dialects include the Ovimbundu, Mbundu, Bakongo, Chokwe, and other peoples.

Heritage

Angola has a rich subsoil heritage, from diamonds, oil, gold, copper. Also a non-commerciable rich wildlife (dramatically empoverished during the civil war), forest, and fossils.

13

The Republic of the Congo (Brazzaville)

The Republic of the Congo, also known as Congo-Brazzaville, Little Congo, or simply the Congo, is a country in Central Africa. It is bordered by Gabon, Cameroon, the Central African Republic, the Democratic Republic of the Congo (formerly known as Zaire), the Angolan exclave province of Cabinda, and the Gulf of Guinea.

The region was dominated by Bantu tribes, who built trade links leading into the Congo River basin. The republic is a former French colony. Upon independence in 1960, the former French region of Middle Congo became the Republic of the Congo. The People's Republic of the Congo was a Marxist-Leninist single-party state from 1970 to 1991. Multiparty elections have been held since 1992, although a democratically elected government was ousted in a 1997 civil war.

History

The earliest inhabitants of the region were Pygmy people, who later were largely displaced and absorbed by Bantu who found tribes during the Bantu expansions. The Bakongo are a Bantu ethnicity that also occupied parts of present-day Angola, Gabon, and Democratic Republic of the Congo, forming the basis for ethnic affinities and rivalries among those countries. Several Bantu kingdoms—notably those of the Kongo, the Loango, and the Teke—built trade links leading into the Congo River basin.

The inhabitants of the Congo river delta first came into contact with Europeans in the late 15th century with Portuguese expeditions charting the African coastline. Commercial relationships were quickly established between the inland Bantu kingdoms and European merchants who traded various commodities, manufactured goods, and slaves captured from the hinterlands. For centuries, the Congo river delta was a major commercial hub for transatlantic trade. However, when direct European colonization of the African continent began in the late 19th century, the power of the Bantu societies in the region eroded.

The area came under French sovereignty in the 1880s. In 1908, France organized French Equatorial Africa (AEF), comprising its colonies of Middle Congo (modern Congo), Gabon, Chad, and Oubangui-Chari (modern Central African Republic). Brazzaville was selected as the federal capital. Economic development during the first 50 years of colonial rule in Congo centred on natural resource extraction. The Conference of 1944 heralded a period of major reform in French colonial policy. Congo benefited from the postwar expansion of colonial administrative and infrastructure spending as a result of its central geographic location within AEF and the federal capital at Brazzaville.

Following independence as the Congo Republic on August 15, 1960, Fulbert Youlou ruled as the country's first president until labour elements and rival political parties instigated a three-day uprising that ousted him. The Congolese military took charge of the country briefly and installed a civilian provisional government headed by Alphonse Massamba-Débat. Under the 1963 constitution, Massamba-Débat was elected President for a five-year term. The regime adopted "scientific socialism" as the country's constitutional ideology.

In 1965, Congo established relations with the Soviet Union, the People's Republic of China, North Korea and North Vietnam. Massamba-Débat was unable to reconcile various institutional and ideological factions and his regime was ended abruptly with an August 1968 coup d'etat. Marien Ngouabi, who had participated in the coup, assumed the presidency on December 31, 1968. One year later, President Ngouabi proclaimed Congo to be Africa's first "people's republic" and announced the decision of the National Revolutionary Movement to change its name to the Congolese Labour Party (PCT). On March 16, 1977, President Ngouabi was assassinated. An 11-member Military Committee of the Party (CMP) was named to head an interim government with Joachim Yhombi-Opango to serve as President of the Republic. Two years later, Yhombi-Opango was forced from power and Denis Sassou Nguesso become the new president.

Sassou Nguesso aligned the country with the Eastern Bloc and signed a twenty-year friendship pact with the Soviet Union. Over the years, Sassou had to rely more on political repression and less on patronage to maintain his dictatorship.

Lissouba, another socialist, did not bring much change. He delayed economic reforms.

Congo's democratic progress was derailed in 1997 when Lissouba and Sassou started to fight over power. As presidential elections scheduled for July 1997 approached, tensions between the Lissouba and Sassou camps mounted. On June 5, President Lissouba's government forces surrounded Sassou's compound in Brazzaville and Sassou ordered members of his private militia (known as "Cobras") to resist. Thus began a four-month conflict that destroyed or damaged much of Brazzaville and caused tens of thousands of civilian deaths. In early October, the Angolan socialist regime began an invasion of Congo to install Sassou to power. In mid-October, the Lissouba government fell. Soon thereafter, Sassou declared himself President.

Controversial elections in 2002 saw Sassou win with almost 90% of the vote cast. His two main rivals Lissouba and Bernard Kolelas were prevented from competing and the only remaining credible rival, Andre Milongo, advised his supporters to boycott the elections and then withdrew from the race. A new constitution, agreed upon by referendum in January 2002, granted the president new powers and also extended his term to seven years as well as introducing a new bicameral assembly. International observers took issue with the organization of the presidential election as well as the constitutional referendum, both of which were reminiscent in their organization of Congo's era of the single-party state. Following the presidential elections, fighting restarted in the Pool region between government forces and rebels lead by Pastor Ntumi; a peace treaty to end the conflict was signed in April 2003.

The regime held the presidential election in July 2009. According to the Congolese Observatory of Human Rights, a non-governmental organisation, the election was marked by "very low" turnout and "fraud and irregularities." The regime announced Sassou as the winner.

Government and Politics

Congo-Brazzaville is an authoritarian regime, according to the Democracy Index. It is ruled by Denis Sassou Nguesso. Internationally, Sassou's socialist regime has been hit by corruption revelations despite attempts to censor them. One French investigation found over 110 bank accounts and dozens of lavish properties in France; Sassou denounced embezzlement investigations as "racist" and "colonial".

Human Rights

As of 2008, most media are owned by the government. There is one government-owned television station, three government-owned radio stations, and three private pro-government radio stations, and a government-owned newspaper.

Many Pygmies in Congo live as slaves to Bantu masters. Now UNICEF and human-rights activists are speaking out. A law that would grant special protections to the Pygmy people is awaiting a vote by the Congo parliament.

Administrative Divisions

The Republic of the Congo is divided into 12 *departements* (departements). Departments are divided into communes and/ or districts. These are:

- Bouenza
- Cuvette
- Cuvette-Ouest
- Kouilou
- Lekoumou
- Brazzaville
- Likouala

- Niari
- Plateaux
- Pool
- Sangha
- Pointe Noire.

Geography and Climate

The Republic of the Congo is located in the western part of Central Africa. Situated on the Equator, it is bordered by the Angola exclave of Cabinda to the south (201 km), the Democratic Republic of the Congo to the south and east (2,410 km), the Central African Republic (467 km) and Cameroon (523 km) to the north and Gabon to the west (1,903 km). Congo has a 169 km long Atlantic coast with several important ports. The Republic of the Congo covers an area of 342,000 km^2, of which 341,500 km^2 is land while 500 km^2 is water. Congo claims 200 nautical miles (370 km) of territorial sea.

The capital of the Republic of the Congo is Brazzaville, located on the Congo river immediately across from Kinshasa, capital of the Democratic Republic of the Congo. With a metropolitan population of approximately 1.5 million, Brazzaville is by far the largest city in the Republic, having almost thrice the population of Pointe-Noire (663,400 as of the 2005 census), the country's second largest city. About 70% of the population lives in Brazzaville, Pointe-Noire, or along the railroad between them.

Environment

Climate

Congo is a tropical nation, which means it has a tropical climate. The wet season lasts from March to June and the dry season for the rest of the year. Temperature and humidity are high as in all tropical nations. The rivers of the country are flooded seasonally.

Ecology

The terrain is a variation of coastal plains, mountaineous regions, plateaus and fertile valleys. About 70 percent of the country's area is covered by rain forest. The highest point, at

1,020 m, is Mont Nabeba in the Mayumbe mountains. The major rivers are the Congo River at the border with the Democratic Republic of the Congo, and the Kouilou-Niari River.

Natural Resources

Natural resources include petroleum, timber, potash, lead, zinc, uranium, copper, phosphates, natural gas, and hydropower. 1.45% of the land is arable, while only 0.15% contains permanent crops. Approximately 10 km^2 is irrigated (1993 estimation).

Environmental Issues

Environmental issues include the high level of air pollution from vehicle emissions, water pollution from the dumping of raw sewage, tap water not being potable, and deforestation.

Congo is party to the international agreements on Biodiversity, Climate Change, Desertification, Endangered Species, Ozone Layer Protection, Tropical Timber 83, Tropical Timber 94, Wetlands. It has signed but not ratified the Law of the Sea.

Congo is located in the central-western part of sub-Saharan Africa, along the Equator. To the south and east of it is the Democratic Republic of Congo. It is also bounded by Gabon to the west, Cameroon and the Central African Republic to the north, and Cabinda (Angola) to the southwest. It has a short Atlantic coast.

The capital, Brazzaville, is located on the Congo River, in the south of the country, immediately across from Kinshasa, the capital of the Democratic Republic of the Congo.

The southwest of the country is a coastal plain for which the primary drainage is the Kouilou-Niari River; the interior of the country consists of a central plateau between two basins to the south and north. Forests are under increasing exploitation pressure.

Since the country is located on the Equator, the climate is consistent year-round, with the average day temperature being a humid 24 °C (75 °F) and nights generally between 16 °C (61 °F) and 21 °C (70 °F). The average yearly rainfall ranges from 1,100 millimetres (43 in) in south in the Niari valley to over 2,000 millimetres (79 in) in central parts of the country. The dry season is from June to August while in the majority of the country the wet season has two rainfall maxima: one in March–May and another in September–November.

In 2006–07, researchers from the Wildlife Conservation Society studied gorillas in heavily forested regions centred on the Ouesso district of the Sangha Region. They suggest a population on the order of 125,000 Western Lowland Gorillas, whose isolation from humans has been largely preserved by inhospitable swamps.

Economy

The economy of the Republic of the Congo is a mixture of village agriculture and handicrafts, an industrial sector based largely on petroleum extraction, support services, and a government characterized by budget problems and overstaffing. The Congo's growing petroleum sector is by far the country's major revenue earner. In the early 1980s, rapidly rising oil revenues enabled the government to finance large-scale development projects with GDP growth averaging 5% annually, one of the highest rates in Africa. However, the government has mortgaged a substantial portion of its oil earnings, contributing to the government's shortage of revenues. The Congolese oil sector is dominated by the French parastatal oil company Elf Aquitaine, which accounts for 70% of the country's annual oil production. In second position is the Italian oil firm Agip. Chevron, independent CMS Nomeco, and Exxon Mobil are among the American companies active in petroleum exploration or production. Following recent discoveries and oil fields currently under development, Congo's oil production is expected to continue to rise significantly in the next few years.

The country's abundant rain forests are the source of timber. Forestry, which led Congolese exports before the discovery of oil, continues to generate 10% of export earnings, although high transportation costs, high wages, and low productivity have hurt the forestry industry in recent years.

Earlier in the decade, Congo's major employer was the state bureaucracy, which had a payroll of 80,000, which is enormous for a country of Congo's size. The World Bank and other international financial institutions pressured Congo to institute sweeping civil service reforms in order to reduce the size of the state bureaucracy and pare back a civil service payroll that amounted to more than 20% of GDP in 1993. The effort to cut back began in 1994 with a 50% devaluation that cut the payroll in half in dollar terms and by a mid-year reduction of nearly 8,000 in civil service employment and resulted in inflation of 61%. Inflation has since subsided.

Between 1994-96, the Congolese economy underwent a difficult transition. The prospects for building the foundation of a healthy economy, however, were better than at any time in the previous 15 years. Congo took a number of measures to liberalize its economy, including reforming the tax, investment, labour, and hydrocarbon codes. Planned privatizations of key parastatals, primarily telecommunications and transportation monopolies, were launched to help improve a dilapidated and unreliable infrastructure. To build on the momentum achieved during the 2-year period, the IMF approved a 3-year ESAF economic program in June 1996.

By the end of 1996, Congo had made substantial progress in various areas targeted for reform. It made significant strides toward macroeconomic stabilization through improving public finances and restructuring external debt. This change was accompanied by improvements in the structure of expenditures, with a reduction in personnel expenditures. Further, Congo benefited from debt restructuring from a Paris Club agreement in July 1996.

This reform program came to a halt, however, in early June 1997 when war broke out. Denis Sassou-Nguesso, who returned to power when the war ended in October 1997, publicly expressed interest in moving forward on economic reforms and privatization and in renewing cooperation with international financial institutions. However, economic progress was badly hurt by slumping oil prices in 1998, which worsened the Republic of the Congo's budget deficit. A second blow was the resumption of armed conflict in December 1998.

Congo's economic prospects remain largely dependent on the country's ability to establish political stability and democratic rule. The World Bank is considering Congo for post-conflict assistance. Priorities will be in reconstruction, basic services, infrastructure, and utilities. President Sassou has publicly expressed interest in moving forward on economic reforms and privatization, as well as in renewing cooperation with international financial institutions. However, the return of armed conflict in 1998 hindered economic reform and recovery.

Congo and the United States ratified a bilateral investment treaty designed to facilitate and protect foreign investment. The country also adopted a new investment code intended to attract foreign capital. Despite this, Congo's investment climate is not considered favourable, offering few meaningful incentives. High costs for labour, energy, raw materials, and transportation; a restrictive labour code; low productivity and high production costs; militant labour unions; and an inadequate transportation infrastructure are among the factors discouraging investment. The recent political instability, war damage, and looting also will undermined investor confidence. As a result, Congo has little American investment outside of the oil sector.

Statistics

GDP - purchasing power parity - $4.15 billion (1999 est.)

GDP - real growth rate: 5% (1999 est.)

GDP - per capita: purchasing power parity - $1,530 (1999 est.)

GDP - composition by sector: *agriculture:* 10%

Industry: 59%

Services: 31% (1997 est.)

Population below poverty line: NA%

Household income or consumption by percentage share:

Lowest 10%: NA%

Highest 10%: NA%

Inflation rate (consumer prices): 4% (1999 est.)

Labour force: NA

Unemployment rate: NA%

Budget:

Revenues: $870 million

Expenditures: $970 million, including capital expenditures of $NA (1997 est.)

Industries: petroleum extraction, cement kilning, lumbering, brewing, sugar milling, palm oil, soap, cigarette making

Industrial production growth rate: NA%

Electricity - production: 503 GWh (1998)

Electricity - production by source:

Fossil fuel: 0.6%

Hydro: 99.4%

Nuclear: 0%

Other: 0% (1998).

Electricity - consumption: 588 GWh (1998).

Electricity - exports: 0 kWh (1998).

Electricity - imports: 120 GWh (1998).

Agriculture - products: cassava (tapioca), sugar, rice, maize, peanuts, coffee, cocoa; forest products.

Exports: $1.7 billion (f.o.b., 1999).

Exports - commodities: petroleum 50%, lumber, plywood, sugar, cocoa, coffee, diamonds, Um Bongo.

Exports - partners: United States 23%, Benelux 14%, Germany, Italy, Taiwan, Mainland China (1998).

Imports: $770 million (f.o.b., 1999).

Imports - commodities: petroleum products, capital equipment, construction materials, foodstuffs.

Imports - partners: France 23%, United States 9%, Belgium 8%, United Kingdom 7%, Italy (1997 est.)

Debt - external: $5 billion (1997).

Economic aid - recipient: $159.1 million (1995).

Currency: 1 Comnunaute Financiere Africaine franc (CFAF) = 100 centimes.

Exchange rates: Communaute Financiere Africaine francs (CFAF) per US$1 – 647.25 (January 2000), 615.70 (1999), 589.95 (1998), 583.67 (1997), 511.55 (1996), 499.15 (1995).

Demographics

The Republic of the Congo's sparse population is concentrated in the southwestern portion of the country, leaving the vast areas of tropical jungle in the north virtually uninhabited. Thus, Congo is one of the most urbanized countries in Africa, with 70% of its total population living in a few urban areas, namely in Brazzaville, Pointe-Noire, or one of the small cities or villages lining the 534-kilometre (332 mi) railway which connects the two cities. In rural areas, industrial and commercial activity has declined rapidly in recent years, leaving rural economies dependent on the government for support and subsistence.

Ethnically and linguistically the population of the Republic of the Congo is diverse—Ethnologue recognises 62 spoken languages in the country—but can be grouped into three categories. The Kongo are the largest ethnic group and form roughly half of the population.

The most significant subgroups of the Kongo are Laari in Brazzaville and Pool regions and Vili around Pointe-Noire and along the Atlantic coast. The second largest group are the Teke who live to the north of Brazzaville with 17% of the population. Boulangui (M'Boshi) live in northwest and in Brazzaville and form 12% of the population.

Before the 1997 war, about 9,000 Europeans and other non-Africans lived in Congo, most of whom were French; only a fraction of this number remains. Around 300 American expatriates reside in the Congo. Nearly 2,000 white South African farmers have expressed interest in going to Congo. Pygmies make up between 5-10% of Congo's population.

The people of Republic of the Congo are largely a mix of Catholics and Protestants, who account for 50.5% and 40.2% of the population respectively.

The majority of Christians in the country are Catholic, while the remaining comprises various other Christian denominations. Followers of Islam make up 1.3% of the population, and this is primarily due to an influx of foreign workers into the urban centres.

Health

Public expenditure on health was at 1.2% of the GDP in 2004, whereas private expenditure was at 1.3%. HIV prevalence is at several percent among 15–49 year olds. Health expenditure was at US$ 30 per capita in 2004 A large proportion of the population is undernourished. There were 20 physicians per 100,000 persons in the early 2000s.

Culture

Education

Public expenditure of the GDP was less in 2002–05 than in 1991. Public education is theoretically free and compulsory for under-16-year olds, but in practice, expenses exist. Net primary enrollment rate was 44% in 2005, much less than the 79% in 1991.

The country has universities. Education between ages six and sixteen is compulsory. Students who complete six years of primary school and seven years of secondary school obtain a baccalaureate.

At the university, students can obtain a bachelor's degree in three years and a master's after four. Marien Ngouabi University—which offers courses in medicine, law, and several other fields—is the country's only public university. Instruction at all levels is in French, and the educational system as a whole models the French system.

The educational infrastructure has been seriously degraded as a result of political and economic crises. There are no seats in most classrooms, forcing children to sit on the floor. Enterprising individuals have set up private schools, but they often lack the technical knowledge and familiarity with the national curriculum to teach effectively. Families frequently enroll their children in private schools only to find they cannot make the payments.

The Republic of the Congo (Leopoldville)

The Republic of the Congo was an independent republic established following the independence granted to the former colony of the Belgian Congo in 1960. The country's post-independence name remained only until 1 August 1964, when it was changed to Democratic Republic of the Congo, to distinguish it from the neighbouring Republic of the Congo, formerly French Congo.

Unrest and rebellion plagued the government until 1965, when Lieutenant General Joseph-Desire Mobutu, by then commander-in-chief of the national army, seized control of the country and declared himself president for five years.

Belgian Congo

The conditions in the Congo improved following the Belgian government's takeover from the Congo Free State. Select Bantu languages were taught in primary schools, a rare occurrence in colonial education. Colonial doctors were to greatly reduce the spread of African trypanosomiasis, commonly known as sleeping sickness.

The colonial administration implemented a variety of economic reforms that focused on the improvement of infrastructure: railways, ports, roads, mines, plantations and industrial areas. The Congolese people, however, lacked political power and faced legal discrimination.

All colonial policies were decided in Brussels and Leopoldville. The Belgian Colony-secretary and Governor-general, neither of whom was elected by the Congolese people, wielded absolute power. Among the Congolese people, resistance against their undemocratic regime grew over time. In 1955, the Congolese upper class (the so-called "evolues"), many of whom had been educated in Europe, initiated a campaign to end the inequality.

During World War II, the small Congolese army achieved several victories against the Italians in North Africa. The Belgian Congo, which was also rich in uranium deposits, supplied the uranium that was used by the United States to build the atomic weapons that were used in the bombings of Hiroshima and Nagasaki in August 1945.

Congo Crisis

In May 1960, the MNC party or Mouvement National Congolais, led by Patrice Lumumba, won the parliamentary elections, and Lumumba was appointed Prime Minister. Joseph Kasavubu, of the ABAKO (Alliance des Bakongo) party, was elected President by the parliament. Other parties that emerged include the Parti Solidaire Africain (or PSA, led by Antoine Gizenga) and the Parti National du Peuple (or PNP led by Albert Delvaux and Laurent Mbariko).

The Belgian Congo achieved independence on 30 June 1960 under the name "Republic of Congo" or "Republic of the Congo" ("Republique du Congo"). As the French colony of Middle Congo

(Moyen Congo) also chose the name "Republic of Congo" upon receiving its independence, the two countries were more commonly known as "Congo-Leopoldville" and "Congo-Brazzaville", after their capital cities.

In 1966, Joseph Mobutu changed the country's official name to "Democratic Republic of the Congo", and in 1971 it was changed again to "Republic of Zaire". Shortly after independence, the provinces of Katanga (with Moise Tshombe) and South Kasai engaged in secessionist struggles against the new leadership.

Subsequent events led to a crisis between President Kasavubu and Prime Minister Lumumba. On 5 September 1960, Kasavubu dismissed Lumumba from office. Lumumba declared Kasavubu's action "unconstitutional" and a crisis between the two leaders developed.

Lumumba had previously appointed Joseph Mobutu chief of staff of the new Congo army, Armee National Congolaise (ANC).

Taking advantage of the leadership crisis between Kasavubu and Lumumba, Mobutu garnered enough support within the army to create sentiment sufficient to inspire mutinous action. With financial support from the United States and Belgium, Mobutu made payments to his soldiers in order to generate their loyalty. The aversion of Western powers towards communism and leftist ideology in general influenced their decision to finance Mobutu's quest to maintain "order" in the new state by neutralizing Kasavubu and Lumumba in a coup by proxy.

On 17 January 1961, Katangan forces, supported by the Belgian government's desire to retain rights to mine for copper and diamonds in Katanga and South Kasai and the U.S. Central Intelligence Agency's desire to remove any leftist sympathizers in the region, assassinated Patrice Lumumba. Through 1960 to 1964 the peacekeeping effort was the largest, most complex, and most costly operation ever carried out by the United Nations. Amidst widespread confusion and chaos, a temporary government led by technicians (College des Commissaires) with Evariste Kimba, and several short governments Joseph Ileo, Cyrille Adoula, Moise Tshombe took over in quick succession.

Coup D'etat

Following five years of extreme instability and civil unrest, Joseph-Desire Mobutu, now Lieutenant General, overthrew Kasavubu in a 1965 Central Intelligence Agency-backed coup. He had the support of the United States on account of his staunch opposition to Communism, which would presumably make him a roadblock to Communist schemes in Africa.

It is also argued that the Western support for Mobutu was also related to his allowing businesses to export the many natural resources of Zaire without worrying about environmental, labour, or other regulations. A one-party system was established, and Mobutu declared himself head of state. He would periodically hold elections in which he was the only candidate.

14

Ghana

The Republic of Ghana is a country located in West Africa. It is bordered by Cote d'Ivoire (Ivory Coast) to the west, Burkina Faso to the north, Togo to the east, and the Gulf of Guinea to the south. The word *Ghana* means "Warrior King" and is derived from the ancient Ghana Empire.

Ghana was inhabited in pre-colonial times by a number of ancient predominantly Akan Kingdoms, including the Akwamu on the eastern coast, the inland Ashanti Empire and various Fante and non-Akan states, like the Ga and Ewe, along the coast and inland. Trade with European states flourished after contact with the Portuguese in the 15th century, and the British established the Gold Coast Crown colony in 1874.

The Gold Coast achieved independence from the United Kingdom in 1957, becoming the first sub-Saharan African nation to do so and the name Ghana was chosen for the new nation to reflect the ancient Empire of Ghana, which once extended throughout much of west Africa. Ghana is a member of the South Atlantic Peace and Cooperation Zone, the Commonwealth of Nations, the Economic Community of West African States, the African Union, and an associate Member of La Francophonie. Ghana is the second largest producer of cocoa in the world and is also home to Lake Volta, the largest artificial lake in the world by surface area. Ghana has one of the best football teams in Africa. In 2009, it became the first African nation to win the FIFA U-20 World Cup by defeating Brazil and, in June 2010, became the third African nation to reach the quarterfinal stage at a FIFA World Cup event.

Etymology

The word *Ghana* means Warrior King and was the title accorded to the kings of the medieval West African Ghana Empire. Geographically, the Ghana Empire was approximately 500 miles (800 km) north and west of modern Ghana, and it ruled territories in the area of the Senegal River and east towards the Niger River, in modern Senegal, Mauritania and Mali.

Ghana was adopted as the legal name for the Gold Coast combined with British Togoland upon gaining independence on March 6, 1957; however, it was not until July 1, 1960 that Ghana asserted its complete autonomy from Britain and became known as the *Republic of Ghana.*

History

There is archaeological evidence which shows that humans have lived in what is present day Ghana from about 1500 BC. Nonetheless, there is no proof that those early dwellers are related to the current inhabitants of the area. Oral tradition has it that many of Ghana's current ethnic groups such as the multi-ethnic Akan, the Ga and the Ewe arrived around the 13th Century AD. However, the Dagomba are believed to be the first settlers, having been fully established by 1210 AD, before the arrival of the other ethnic groups.

Modern Ghanaian territory includes what was the Empire of Ashanti, one of the most influential states in sub-Saharan Africa before colonial rule. Akan migrants moved southward and founded several nation-states including the first great Akan empire of the Bono, which is now known as the Brong-Ahafo region in Ghana. Much of the area of modern day south central Ghana was united under the Empire of Ashanti of the Ashanti people, a branch of the Akan people by the 16th century.

The Ashanti government operated first as a loose network and eventually as a centralized kingdom with an advanced, highly specialized bureaucracy centred in Kumasi. It is said that at its peak, the Asantehene could field 500,000 troops and had some degree of military influence over all of its neighbours. Among the Ashanti a third of the population were slaves. The Ga people developed an effective unit around 1500 and the

Gonja, Dagomba and Mamprusi also fought for political power in the 1620s.

Early European contact by the Portuguese, who came to Ghana in the 15th century, focused on the extensive availability of gold. The Portuguese first landed at a coastal city inhabited by the Fante nation-state and they named the place *Elmina*, which means "the mine" in Portuguese. In 1481, King John II of Portugal commissioned Diogo d'Azambuja to build Elmina Castle, which was completed in 3 years. Their aim was to trade in gold, ivory and slaves, consolidating their burgeoning political and economic power in the region.

By 1548, the Dutch had joined them, and built forts at Komenda and Kormantsi. In 1617, they captured the Olnini Castle from the Portuguese and Axim in 1642 (Fort St Anthony). Other European traders joined in by the mid 17th century, largely English, Danes and Swedes. British merchants, impressed with the gold resources in the area, named it the *Gold Coast*, while French merchants, impressed with the trinkets worn by the coastal people, named the area to the west "Cote d'Ivoire", or Ivory Coast.

More than thirty forts and castles were built by the Portuguese, Dutch, British and Spanish merchants. The Gold Coast was known for centuries as 'The White Man's Grave' because many of the Europeans who went there died of malaria and other tropical diseases. After the Dutch withdrew in 1874, Britain made the Gold Coast a protectorate. Following conquest by the British in 1896, until independence in March 1957, the territory of modern Ghana excluding the Volta Region (British Togoland), was known as the Gold Coast.

Many wars occurred between the colonial powers and the various nation-states in the area including the 1806 Ashanti-Fante War and the continuous struggle by the Ashanti against the British, which ended in 1901 with the Third Ashanti-British War (1900–1901). Even under colonial rule the chiefs and people often resisted the policies of the British; however, moves toward de-colonization intensified after World War II. In 1947 the newly formed United Gold Coast Convention (UGCC) called for "self-government within the shortest possible time." After rioting increased in 1948, the members of the United

Gold Coast Convention were arrested, including future Prime Minister and President, Kwame Nkrumah. Later Nkrumah formed his own party, the Convention People's Party (CPP) with the motto "self government now." He began a 'Positive Action' campaign and gained the support of rural and working class people.

Once again he was imprisoned for being the leader of a party that caused boycotts, strikes and other forms of civil disobedience. After winning a majority of seats in the Legislative Assembly in 1952; however, Kwame Nkrumah was released and appointed Leader of Government Business. After further negotiations with Britain finally on March 6, 1957 at 12 a.m. Kwame Nkrumah declared Ghana "free forever".

The flag which consists of the colours red, gold, green and the black star became the new flag in 1957. Designed by Theodosia Salome Okoh, the red represents the blood that was shed towards independence, the gold represents the mineral wealth of Ghana, the green symbolises the rich agriculture and the black star is the symbol of African emancipation.

Formed from the merger of the Gold Coast and British Togoland by a United Nations sponsored plebiscite in 1956, Ghana became the first sub-Saharan African country to gain its independence in 1957.

Kwame Nkrumah, first Prime Minister and then President of the modern Ghanaian state, was not only an African anti-colonial leader but also one with a dream of a united Africa which would not drift into neo-colonialism. He was the first African head of state to promote Pan-Africanism, an idea he came into contact with during his studies at Lincoln University in Pennsylvania (United States), at the time when Marcus Garvey was becoming famous for his "Back to Africa Movement." He merged the dreams of both Marcus Garvey and the celebrated African-American scholar W. E. B. Du Bois into the formation of the modern day Ghana. Ghana's principles of freedom and justice, equity and free education for all, irrespective of ethnic background, religion or creed, borrow from Kwame Nkrumah's implementation of Pan-Africanism.

Although his goal of African unity never realised, Osagyefo Dr. Kwame Nkrumah, as he is now known, played an instrumental part in the founding of the Organisation of African Unity, which was succeeded in 2002 by the African Union. His achievements were recognised by Ghanaians during his Centenary birthday celebrations and the day instituted as a public holiday in Ghana. Dr. Kwame Nkrumah's government was subsequently overthrown by the military while abroad in February 1966. It is believed by many political analysts that the United States' Central Intelligence Agency (CIA) participated in the coup, but that generally remains unproven.

A series of subsequent coups from 1966 to 1981 ended with the ascension to power of Flight Lieutenant Jerry Rawlings in 1981. These changes resulted in the suspension of the constitution in 1981 and the banning of political parties. The economy suffered a severe decline soon after, and many Ghanaians migrated to other countries. Although most migrating Ghanaians went to Nigeria, the Nigerian government deported about a million Ghanaians back to Ghana in 1983.

Jerry Rawlings soon negotiated a structural adjustment plan with the International Monetary Fund and changed many old economic policies and, thus, the economy soon began to recover. A new constitution restoring multi-party politics was promulgated in 1992, and Rawlings was elected as president then and again in 1996. The Constitution of 1992 prohibited him from running for a third term, so his party, the National Democratic Congress, chose his Vice President, John Atta Mills, to run against the opposition parties. Winning the 2000 elections, John Kufuor of the New Patriotic Party was sworn into office as President in January 2001, and beat Mills again in 2004; thus, also serving two terms as President.

Since 2008 Ghana have had several issues with its taxing and income of most of its workers. It has had several issues with ongoing security which has led to a major decrease in the country's overall wealth. As well as this the next president promised to build more tarmac roads as he commented on its major lack of them.

In 2009, John Atta Mills took office as president with a difference of about 40,000 votes (0.46%) between his party, the National Democratic Congress, and the New Patriotic Party, marking the second time that power had been transferred from one legitimately elected leader to another, and securing Ghana's status as a stable democracy.

Regions and Districts

Ghana is divided into 10 administrative regions, subdivided into a total of 138 districts. The regions are:

- Ashanti, capital Kumasi
- Brong Ahafo, capital Sunyani
- Central, capital Cape Coast
- Eastern, capital Koforidua
- Greater Accra, capital Accra
- Northern, capital Tamale
- Upper East, capital Bolgatanga
- Upper West, capital Wa
- Volta, capital Ho
- Western, capital Sekondi-Takoradi

Population of Major Cities

City	*Population*
Accra	3,963,264
Kumasi	2,604,909
Tamale	390,730
Takoradi	260,651
Tema	229,106
Teshie	154,513
Sekondi	153,900
Cape Coast	200,204
Obuasi	147,613
Dunkwa-On-Offin	108,482

Government and Politics

According to the 2009 Failed States Index, Ghana is ranked the 53rd least failed state in the world and the second least failed state in Africa after Mauritius. Ghana ranked 124th out of 177 countries on the index. Ghana also was placed 7th out of 48 sub-Saharan African countries in the 2008 Ibrahim Index of African Governance which was based on data from 2006. The Ibrahim Index is a comprehensive measure of African government, based on a number of different variables which reflect the success with which governments deliver essential political goods to its citizens.

Government: Ghana was created as a parliamentary democracy at independence in 1957, followed by alternating military and civilian governments. In January 1993, military government gave way to the Fourth Republic after presidential and parliamentary elections in late 1992. The 1992 constitution divides powers among a President, Parliament, Cabinet, Council of State, and an independent judiciary. The Government is elected by universal suffrage; however, the legislature is greatly malapportioned, with low-population districts receiving more representatives per person than those with high populations.

Administrative Divisions: There are ten administrative regions which are divided into 138 districts, each with its own District Assembly. Below districts are various types of councils, including 58 town or area councils, 108 zonal councils, and 626 area councils. 16,000 unit committees on lowest level.

Judicial System: The legal system is based on British common law, customary (traditional) law, and the 1992 constitution. Court hierarchy consists of Supreme Court of Ghana (highest court), Courts of Appeal, and High Courts of Justice. Beneath these bodies are circuit, magisterial, and traditional courts. Extrajudicial institutions include public tribunals. Since independence, courts are relatively independent; this independence continues under Fourth Republic. Lower courts are being redefined and reorganized under the Fourth Republic.

Politics: Political parties became legal in mid-1992 after a ten-year hiatus. There are many political parties under the Fourth Republic; however, the major ones are the National Democratic Congress which won presidential and parliamentary elections in 1992, 1996 and 2008; the New Patriotic Party, the major opposition party which won elections in 2000 and 2004; the People's National Convention, and the Convention People's Party, successor to Kwame Nkrumah's original party of the same name. Foreign Relations: Since independence, Ghana has been fervently devoted to ideals of nonalignment and Pan-Africanism, both closely identified with first president, Osagyefo Dr. Kwame Nkrumah. Ghana favors international and regional political and economic co-operation, and is an active member of the United Nations and the African Union.

Many Ghanaian diplomats and politicians hold positions in international organisations. These include Ghanaian diplomat and former Secretary-General of the United Nations, Kofi Annan, International Criminal Court Judge Akua Kuenyehia, and former president Jerry Rawlings, who was elected chairman of the Economic Community of West African States.

Economy

Well endowed with natural resources, Ghana has twice the per capita output of the poorer countries in West Africa. Even so, Ghana remains somewhat dependent on trade and international assistance as well as the investment activities of Ghanaian diaspora. About 28% of the population live below the international poverty line of US$1.25 a day, the vast majority of which are Ghanaian women from the politically marginalised and poor northern and upper regions and according to the World Bank, Ghana's per capita income has barely doubled over the past 45 years. Ghana, known for its gold in colonial times, remains one of the world's top gold producers. Other exports such as cocoa, timber, electricity, diamond, bauxite, and manganese are major sources of foreign exchange monitored, operated and managed by the Presidential Ministry Agricultural Arm of the Republic of Ghana headed by Mrs. Antoinette Efua-Addo. An oilfield which is reported to contain up to 3 billion barrels (480,000,000 m^3) of light oil was discovered in 2007. Oil exploration is ongoing and the amount of oil continues to increase. There is expected to be a tremendous inflow of capital into the economy beginning from the last quarter of 2010 when the country starts producing oil in commercial quantities.

The Akosombo Dam, which was built on the Volta River in 1965 provides hydro-electricity for Ghana and its neighbouring countries. Ghana's labour force in 2008 totalled 11.5 million people The economy continues to rely heavily on agriculture which accounts for 37.3% of GDP and provides employment for 56% of the work force, mainly small landholders. Manufacturing is only a small part of the Ghanaian economy totalling 7.9% of Gross Domestic Product in 2007.

Ineffective economic policies of past military governments and regional peacekeeping commitments have led to continued inflationary deficit financing, depreciation of the Cedi, and rising public discontent with Ghana's austerity measures. Even so, Ghana remains one of the more economically sound countries in all of Africa.

In July 2007, the Bank of Ghana embarked on a currency re-denomination exercise, from the Cedi to the new currency, the Ghana Cedi. The transfer rate is 1 Ghana Cedi for every 10,000 Cedis. The Bank of Ghana employed aggressive media campaigns to educate the public about the re-denomination.

The new Ghana Cedi is relatively stable and in 2009 generally exchanged at a rate of $1 USD =Gh¢ 1.4 The Value Added Tax is a consumption tax administered in Ghana. The tax regime which started in 1998 had a single rate but since September 2007 entered into a multiple rate regime.

In 1998, the rate of tax was 10% and amended in 2000 to 12.5%. However with the passage of Act 734 of 2007, a 3% VAT Flat Rate Scheme (VFRS) began to operate for the retail distribution sector. This allows retailers of taxable goods under Act 546 to charge a marginal 3% on their sales and account on same to the VAT Service. It is aimed at simplifying the tax system and increasing compliance.

Geography

Ghana is a country located on the Gulf of Guinea, only a few degrees north of the Equator, therefore giving it a warm climate. The country spans an area of 238,500 km^2 (92,085 sq mi). It is surrounded by Togo to the east, Cote d'Ivoire to the west, Burkina Faso to the north and the Gulf of Guinea (Atlantic Ocean) to the south. The Greenwich Meridian passes through Ghana, specifically through the industrial city of Tema. Ghana is geographically closer to the "centre" of the world than any other country even though the actual centre, (0°, 0°) is located in the Atlantic Ocean approximately 614 km (382 mi) south of Accra, Ghana, in the Gulf of Guinea. The country encompasses flat plains, low hills and a few rivers. Ghana can be divided into five different geographical regions. The coastline is mostly a low, sandy shore backed by plains and scrub and intersected by several rivers and streams while the northern part of the country features high plains. Southwest and south central Ghana is made up of a forested plateau region consisting of the Ashanti uplands and the Kwahu Plateau and the hilly Akuapim-Togo ranges are found along the country's eastern border.

The Volta Basin also takes up most of central Ghana. Ghana's highest point is Mount Afadjato which is 885 m (2,904 ft) and is found in the Akwapim-Togo Ranges. The climate is tropical. The eastern coastal belt is warm and comparatively dry; the southwest corner, hot and humid; and the north, hot and dry. Lake Volta, the world's largest artificial lake, extends through large portions of eastern Ghana and is the main source of many tributary rivers such as the Oti and Afram rivers.

There are two main seasons in Ghana; the wet and the dry seasons. Northern Ghana experiences its rainy season from March to November while the south, including the capital Accra, experiences the season from April to Mid-November. Southern Ghana contains evergreen and semideciduous forests consisting of trees such as mahogany, odum and ebony. It also contains much of Ghana's oil palms and mangroves. Shea trees, baobabs and acacias are usually found in the Volta region and the northern part of the country.

Demographics

Ghana has a population of about 24 million people. It is home to more than 100 different ethnic groups. Ghana has not seen the kind of ethnic conflict that has created civil wars in many other African countries. The official language is English; however, most Ghanaians also speak at least one local language.

The ethnic groups in Ghana are the Akan (which includes the Fante, Akyem, Ashanti, Kwahu, Akuapem, Nzema, Bono, Akwamu, Ahanta and others) 49.3%, Mole-Dagbon 15.2%, Ewe 11.7%, Ga-Dangme (comprising of the Ga, Adangbe, Ada, Krobo and others) 7.3%, Guan 4%, Gurma 3.6%, Gurunsi 2.6%, Mande-Busanga 1%, other tribes 1.4%, other (Hausa, Zabarema, Fulani)

1.8% (2000 census). According to the CIA World Factbook, religious divisions are as follows: Christian 68.8%, Muslim 15.9%, Traditional African beliefs 8.5%.

Health

As of 2009, life expectancy at birth is about 59 years for males and 60 years for females with infant mortality at 51 per 1000 live births. The birth rate is also about 4 children born per woman. There are about 15 physicians and 93 nurses per 100,000 persons. 4.5% of the country's GDP was spent on health in 2003.

Languages

Ghana has 47 local languages. English is the country's official language and predominates government and business affairs. It is also the standard language used for educational instruction. Native Ghanaian languages are divided into two linguistic subfamilies of the Niger-Congo language family. Languages belonging to the Kwa subfamily are found predominantly to the south of the Volta River, while those belonging to the Gur subfamily are found predominantly to the north. The Kwa group, which is spoken by about 75% of the country's population, includes the Akan, Ga-Dangme, and Ewe languages. The Gur group includes the Gurma, Grusi, and Dagbani languages.

Nine languages have the status of government-sponsored languages: Akan, specifically Ashanti Twi, Fanti, Akuapem Twi, Akyem, Kwahu, Nzema; Dagaare/Wale, Dagbani, Dangme, Ewe, Ga, Gonja and Kasem. Though not an official language, Hausa is the lingua-franca spoken among Ghana's Muslims who comprise about 16% of the population.

People and culture

Ghana is an ethnically diverse country; thus, Ghanaian culture is a mixture of all its ethnic groups, the Ashanti, Fante, Akyem, Kwahu, Ga, Ewe, Mamprusi and Dagomba, among others. It is most evident in Ghanaian cuisine, the arts and clothing. The celebration of festivals in Ghana is an essential part of Ghanaian culture and there are many of them such as the Homowo, Odwira, Aboakyer, Dodoleglime, Hogbetsotso,

Tedudu, Deza (festival) and Sandema among others. Several rites and rituals are performed throughout the year in various parts of the country, including child-birth, rites of passage, puberty, marriage and death.

Sports

Association Football is the most popular sport in the country. The national men's football team is known as the Black Stars, with the under-20 team known as the Black Satellites. The under 17s are the Black Starlets. They have participated in many championships including the African Cup of Nations, the FIFA World Cup and the FIFA U-20 World Cup. On October 16, 2009, Ghana became the first African nation to win the FIFA U-20 World Cup by defeating Brazil 4-3 in a penalty shoot-out. On June 13, 2010, Ghana defeated Serbia 1-0 in first round play in the 2010 FIFA World Cup becoming the first African team to win a FIFA World Cup finals game on African soil and subsequently became the only African team to progress from the group stage to the knock out phase at the 2010 event. On June 26 Ghana reached the quarter final stage after defeating USA 2-1 in their last 16 knock out match, but they fell short of reaching the semifinals after being defeated in a penalty shoot-out by Uruguay in Johannesburg on July 2, 2010. There are several football teams in Ghana more notably the Accra Hearts of Oak SC and Asante Kotoko among others. Some Ghanaian football players that are recognised on an international level or achieved success in European football are Michael Essien, Abedi Pele, Asamoah Gyan, Ibrahim Abdul Razak, Tony Yeboah, Anthony Annan, Quincy Owusu-Abeyie, John Paintsil, Kevin-Prince Boateng, Samuel Osei Kuffour, Richard Kingson, Sulley Muntari, Laryea Kingston, Stephen Appiah, Andre Ayew, Emmanuel Agyemang-Badu, John Mensah and Dominic Adiyiah. Ghana is also the birth place of World Wrestling Entertainment Wrestler Kofi Kingston (born Kofi Sarkodie-Mensah), who is wrestling on the Smackdown brand. Also is Kwame Nkrumah-Acheampong who competed in the Vancouver Winter Olympics. The country has also produced quite a few quality boxers such as Azumah Nelson a three time world champion, Nana Yaw Konadu also a three time world champion, Ike Quartey, and Joshua Clottey, who lost his most recent fight to boxing champion Manny Pacquiao.

Fashion

Textiles are very important in Ghanaian culture. These cloths are used to make traditional and modern attire. Different symbols and different colours mean different things. The Kente is probably the most famous of all the Ghanaian cloths. *Kente* is an Ashanti ceremonial cloth hand-woven on a horizontal treadle loom. Strips measuring about 4 inches wide are sewn together into larger pieces of cloths. Cloths come in various colours, sizes and designs and are worn during very important social and religious occasions.

In a cultural context, kente is more important than just a cloth. It is a visual representation of history, and also form of a written language through weaving. The term kente has its roots in the Twi word *kenten* which means a basket. The first kente weavers used raffia fibers to weave cloths that looked like kenten (a basket); and thus were referred to as kenten ntoma; meaning basket cloth. The original Asante name of the cloth was *nsaduaso* or *nwontoma*, meaning "a cloth hand-woven on a loom"; however, the term kente is the most popularly used term today. Many variations of narrow-strip cloths similar to *kente* are woven by various ethnic groups in Ghana like the Ewe, Ga and others in Africa. It is also popular among the African diaspora.

Fine Arts

Ghana has been recognized on the international level through several artists, including the Kane Kwei Carpentry Workshop and Eric Adjetey Anang who are creating the famous design coffins.

Music

Ghana has many types of traditional and modern music. The sound varies from ethnic group to ethnic group and region to region. Ghanaian music incorporates several distinct types of musical instruments such as the talking drum ensembles, goje fiddle and koloko lute, court music, including the Akan atumpan, the Ga kpanlogo styles, and log xylophones used in asonko music. The most well known genres to have come from Ghana are Afro-jazz which was created by Ghanaian artist Kofi Ghanaba and its earliest form of secular music is called Highlife.

Highlife originated in the late 1800 and early 1900s and spread throughout West Africa mainly Sierra Leone and Nigeria. In the 1990's a new genre of music was created by the youth incorporating the influences of Highlife Afro-reggae, Dancehall and Hiphop. This hybrid was called Hiplife. Ghanaian artists such as R&B and Soul singer Rhian Benson and Highlife singer Kojo Antwi have had international success.

Dance

Ghanaian dance is as diverse as its music. Each ethnic group has their own traditional dances and there are different dances for different occasions. There are dances for funerals, celebrations, storytelling, praise and worship etc. Some of these dances include Bamaya It is performed by the Northern people of Ghana. It narrates the legend of a time of great drought. An oracle told the people that the drought was brought about by the manner in which the men were severely repressing and demeaning the women. It further stated that the drought would be relieved only when the men lowered themselves to the role they were imposing on the women by putting on skirts and participating in this dance. When the men did this it began to rain. It is currently performed during harvest time in northwestern Ghana by both Dagbani men and women.

Adowa A dance of the Ashanti peoples of Ghana. This dance is especially noted for the grace and complexity of the dancers' movements. The drumming is also noted for the complexity of the interlocking rhythms and the two atumpan drums which are used as the lead or master drum. Originally funeral dance music, Adowa is now also performed at annual festivals and social gatherings.

Kpanlongo Is performed by the Ga people of Ghana. It is often referred to as "the dance of the youth," Kpanlongo started during the wake of Ghana's Independence as a musical type for entertainment in Accra. Kpanlongo is presently performed at life-cycle events, festivals, and political rallies.

Klama Is the music and dance is associated with puberty rites of the Krobo people of Ghana. It emphasizes the graceful movement of hands and feet. With small rhythmic steps and heads turned demurely downward, the dancers embody quiet elegance. The different movements of the dance are designed to reveal the beauty of the dancers. Suitors watching from the sidelines will often approach a girl's family after the ceremony and make an offer for her hand in marriage.

Agbadza The traditional dance of the E‹e (Ewe or Eve) people of Ghana. It is characterized by the graceful choreograph of a couple seasoned with the rhythmic movement of the arms, the waist and the feet in perfect synchrony. Agbadza, is traditionally a war dance but is now used in social and recreational situations to celebrate peace. War dances are sometimes used as military training exercises, with signals from the lead drum ordering the warriors to move ahead, to the right, go down, etc. These dances also helped in preparing the warriors for battle and upon their return from fighting they would act out their deeds in battle through their movements in the dance. Atsiagbekor is a contemporary version of the Ewe war dance Atamga (Great (ga) Oath (atama) in reference to the oaths taken by people before proceeding into battle. The movements of this present-day version are mostly in platoon formation and are not only used to display battle tactics, but also to energize and invigorate the soldiers. Today, Atsiagbekor is performed for entertainment at social gatherings and at cultural presentations.

Atsia dance is performed mostly by women, and its a series of stylistic movements dictated to dancers by the lead drummer. Each dance movement has its own prescribed rhythmic pattern, which is synchronized with the lead drum. 'Atsia' in the Ewe language means style or display.

BTbTTbT (pronounced Borborbor) the Ewe-speaking people in the central and northern parts of the Volta Region of Ghana cultivate the BTbTTbT dance. BTbTTbT (originally 'Akpese') might have originated in the Kpando area, and is said to have been created by the late Mr. Francis Kojo Nuadro. He is thought to have been an ex-police officer who returned to Kpando and organized a group in the middle to late 1940's. The dance has its roots in the 'Highlife' popular music of Ghana and other West African countries. BTbTTbT gained national recognition in the 1950's and 1960's because of its use at political rallies and the novelty of its dance formations and movements. It is generally performed at funerals and other social occasions. This is a social dance with a great deal of room for free expression. In general, the men sing and dance in the centre while the women dance in a ring around them. There are 'slow' and 'fast' versions of BTbTTbT; the fast BTbTTbT is believed to come from the Kpando area and the slow version from Hohoe. The slow one is called Akpese and the fast one is termed to be BTbTTbT. Lolobi-Kumasi is known for doing a particular fast version of the slow version.

Agahu is both the name of a dance and of one the many secular music associations (clubs) of the Ewe people of Ghana, Togo, and Dahomey. (Gadzok, Takada, and Atsiagbeko are other such clubs). Each club has its own distinctive drumming and dancing, as well as its own repertoire of songs. A popular social dance of West Africa, Agahu was created by the Egun speaking people from the town of Ketonu in what is now Benin. From there it spread to the Badagry area of Nigeria where migrant Ewe fisherman heard, adapted, and eventually took it to Ghana. In dancing the Agahu, two circles are formed; the men stay stationary with their arms out and then bend with a knee forward for the women to sit on. They progress around the circle until they arrive at their original partner.

Religion

According to the CIA World Factbook, religious divisions are as follows: Christian 68.8%, Muslim 15.9%, Traditional African beliefs 8.5%.

Media and Entertainment

The media of Ghana is one of the most free in Africa, and had previously undergone a series of government overthrows by military leaders and periods of severe restriction. Chapter 12 of the 1992 Constitution of Ghana guarantees freedom of the press and independence of the media, while Chapter 2 prohibits censorship.

Post independence, the government and media often had a tense relationship, with private outlets closed during the military coups and strict media laws that prevent criticism of government. The media freedoms were restored in 1992, and after the election in 2000 of John Kufuor the tensions between the private media and government decreased. Kufuor was a supporter of press freedom and repealed a libel law, though maintained that the media had to act responsibly. The Ghanaian media has been described as "one of the most unfettered" in Africa, operating with little restriction on private media. The private press often carries criticism of government policy. The media were vigorous in their coverage of the 2008 Ghanaian presidential election, and the Ghanaian Journalists Association (GJA) praised John Atta Mills on his election, hoping to foster a good media-government relationship.

Education

The adult literacy rate in Ghana was 65% in 2007 , with males at 71.7% and females at 58.3%. Ghana has a 6-year primary education system beginning at the age of six, and, under the educational reforms implemented in 1987 and reformed in 2007, they pass on to a 3-year junior high school system. At the end of the 3rd year of Junior High, there is a mandatory Basic Education Certificate Examination (BECE). Those continuing must complete the 3-year senior high school (SHS) program and take an admission exam to enter any university or tertiary programme.

Presently, Ghana has 21,530 primary schools, 8,850 junior secondary schools, 900 senior secondary schools, 52 public training colleges, 5 private training colleges, 5 polytechnical institutions, 4 non-university public tertiary institutions, 8 public universities and over 45 private tertiary institutions. Most Ghanaians have relatively easy access to primary and secondary education. These numbers can be contrasted with the single university and handful of secondary and primary schools that existed at the time of independence in 1957. Ghana's spending on education has varied between 28 and 40 percent of its annual budget in the past decade. All teaching is done in English, Ghana's official language, mostly by qualified Ghanaian educators.

The courses taught at the Primary or Basic School level include English, Ghanaian language and Culture, Mathematics, Environmental studies, Social Studies and French as a Third language are added, Integrated or General Science, Pre-vocational Skills and Pre-technical skills, Religious and Moral Education, and physical activities such as Music, Dance and Physical Education. The Senior High level School curriculum has Core subjects and Elective subjects of which students must take four the core subjects of English language, Mathematics, Integrated Science (including Science, Agriculture and Environmental studies) and Social Studies (economics, geography, history and government).

The High school students also choose 3 elective subjects from 5 available programmes: Agriculture Programme, General Programme (Arts or Science option), Business Programme, Vocational Programme and Technical programme. Apart from most primary and secondary schools which choose the Ghanaian system of schooling, there are also international schools such as the Ghana International school, The Roman Ridge School, the Lincoln Community School and the Gmeiner International College, which offer the International Baccalaureat, Advanced Level General Certificate of Education and the International General Certificate of Secondary Education (IGCSE).

With 83 percent of its children in school, Ghana currently has one of the highest school enrolment rates in West Africa. The ratio of girls to boys in the total education system is 1:0.96, which for a West African country, is a considerable achievement. That said, some 500,000 children still remain out of school because of resource constraints in building schools, providing adequate textbooks and training new teachers. UNESCO reports that sixth-graders sitting a simple multiple-choice reading test scored on average the same mark that would be gained by random guessing. The oldest university in Ghana, the University of Ghana, which was founded in 1948, had a total of about 29,754 students in 2008. Since Ghana's independence, the country has been one of the educational hot spots in Sub-Saharan Africa and has played host to notables such as President Robert Mugabe of Zimbabwe, Alhaji Sir Dauda Jawara of The Gambia and Cyprian Ekwensi of Nigeria among others. Former UN Secretary General Kofi Annan has been chancellor of the University of Ghana since 2008.

Kwame Nkrumah University of Science and Technology, the second university to be established in Ghana, is the premier university of science and technology in Ghana and the West Africa sub region.

Bibliography

Belden C.: *Landscapes of the Sacred: Geography and Narrative in American Spirituality*, New York: Paulist Press, 1988.

Berry, J.K. : *Beyond Mapping, Concepts, Algorithms and Issues in GIS*, Fort Collins, CO: GIS World Books, 1993.

Bolstad, P. : *GIS Fundamentals, A first text on Geographic Information Systems, Second Edition*, White Bear Lake, MN, Eider Press, 2005.

Cronon, William: *Changes in the Land: Indians, Colonists, and the Ecology of New England*. New York: Hill and Wang, 1983.

Davis, Mike: *Ecology of Fear: Los Angeles and the Imagination of Disaster*. New York: Vintage, 1999.

Elangovan K. : *GIS, Fundamentals, Applications and Population Geography*, New India Publishing Agency, New Delhi 2006.

Harvey, David: *Explanation in Geography*. London: Edward Arnold, 1969.

Harvey, Francis : *A Primer of GIS, Fundamental Geographic and Cartographic Concepts,* The Guilford Press, 2008.

Jackson, Peter: *Maps of Meaning: An Introduction to Cultural Geography*. London and Boston: Unwin Hyman, 1989.

James S. Duncan: *Writing Worlds: Discourse, Text, and Metaphor in the Representation of Landscape*. London and New York: Routledge, 1992.

James, T.: *The Best Poor Man's Country: A Geographical Study of Early Southeastern Pennsylvania*. Baltimore: Johns Hopkins Press, 1972.

King, Anthony D.: *Buildings and Society: Essays on the Social Development of the Built Enviroment*. London: Routledge & Kegan Paul, 1980.

Lewis, Peirce. *New Orleans: The Making of an Urban Landscape.* Cambridge, MA: Ballinger Publishing Co., 1976. Annotation

Liam Kennedy: *Urban Space and Representation.* Sterling, VA: Pluto Press, 2000.

Lung, Y., and van Tulder, R. : *Cars, Carriers of Regionalism*, Palgrave Macmillan, New York, 2004.

McDonnell, R.A. : *Principles of Geographical Information Systems*, Oxford University Press, Oxford, 1998.

Michael P.: *The Making of the American Landscape.* Boston: Unwin Hyman, 1990.

Michael, H.: *City Form and Natural Process: Towards a New Urban Vernacular.* London: Croom Helm, 1984.

Peter, G.: *The Slow Plague: A Geography of the AIDS Pandemic.* Oxford and Cambridge, MA: Blackwell Publishers, 1993.

Ray Allen: *The Far Western Frontier, 1830-1860.* The New American Nation Series. New York: Harper, 1956.

Richard SCASE : *Managing Creativity,* Open University Press, 2000.

Stephen, D.: *Fields of Vision: Landscape Imagery and National Identity in England and the United States.* Cambridge: Polity, 1993.

Tulder, R. : *Cars, Carriers of Regionalism*, Palgrave Macmillan, New York, 2004.

Victor, A. : *Geography of Art and Culture*, Contributions to Economic Analysis, Elsevier, 2004.

Watkins, Eric: *The Middle Eastern Environment*, London, The British Society for Middle Eastern Studie, 1995.

William G. BOWEN : *Performing Arts, The Economic Dilemma,* Thousand Oaks, SAGE, London, 2002.

Worboys, Michael, and Matt Duckham. : *GIS, a Computing Perspective*, Boca Raton, CRC Press, 2004.

Index

F

G

H

I

J

L

M

N

O

P

□□□